A Legal Primer for the Digital Age

THE ALLYN AND BACON SERIES IN TECHNICAL COMMUNICATION

Series Editor: Sam Dragga, Texas Tech University

Thomas T. Barker
Writing Software Documentation: A Task-Oriented Approach, Second Edition

Carol M. Barnum
Usability Testing and Research

Deborah S. Bosley
Global Contexts: Case Studies in International Technical Communication

Melody Bowdon and Blake Scott
Service-Learning in Technical and Professional Communication

R. Stanley Dicks
Management Principles and Practices for Technical Communicators

Paul Dombrowski
Ethics in Technical Communication

David Farkas and Jean Farkas
Principles of Web Design

Laura J. Gurak
Oral Presentations for Technical Communication

Sandra W. Harner and Tom G. Zimmerman
Technical Marketing Communications

TyAnna K. Herrington
A Legal Primer for the Digital Age

Richard Johnson-Sheehan
Writing Proposals: Rhetoric for Managing Change

Dan Jones
Technical Writing Style

Charles Kostelnick and David D. Roberts
Designing Visual Language: Strategies for Professional Communicators

Victoria M. Mikelonis, Signe T. Betsinger, and Constance Kampf
Grant Seeking in an Electronic Age

Ann M. Penrose and Steven B. Katz
Writing in the Sciences: Exploring Conventions of Scientific Discourse, Second Edition

Carolyn Rude
Technical Editing, Third Edition

Gerald J. Savage and Dale L. Sullivan
Writing a Professional Life: Stories of Technical Communicators On and Off the Job

A Legal Primer
for the Digital Age

TyAnna K. Herrington
Georgia Institute of Technology

PEARSON
Longman

New York   San Francisco   Boston
London   Toronto   Sydney   Tokyo   Singapore   Madrid
Mexico City   Munich   Paris   Cape Town   Hong Kong   Montreal

Senior Vice President and Publisher: Joe Opiela
Vice President and Publisher: Eben W. Ludlow
Executive Marketing Manager: Ann Stypuloski
Senior Supplements Editor: Donna Campion
Media Supplements Editor: Nancy Garcia
Production Manager: Charles Annis
Project Coordination, Text Design, and Electronic Page Makeup: Shepherd Incorporated
Cover Designer/Manager: John Callahan
Manufacturing Buyer: Roy Pickering
Printer and Binder: Hamilton Printing Co.
Cover Printer: Lehigh Press, Inc.

Library of Congress Cataloging-in-Publication Data
On file with the Library of Congress.

Please visit us at http://www.ablongman.com/herrington

ISBN 0-321-10873-6

1 2 3 4 5 6 7 8 9 10—HT—06 05 04 03

For my sister, Kristen Herrington, a very creative communicator

CONTENTS

List of Figures   xi
Preface   xiii

CHAPTER 1

Introduction
1

Discussion Questions and Exercises   6

CHAPTER 2

Law and Ethics
7

Ethics Contextual Situations   19

Discussion Questions   23

CHAPTER 3

Business Relationships and Organizational Structures
25

**Your Characterization in Business Relationships
and Its Legal Effects   26**
Agency Relationships   26
How to Create an Agency Relationship   28

Agency-Principal Relationships and Mutual Duties   29
Three Kinds of Principals   29
Different Kinds of Agents   29

Principals' and Agents' Duties and Obligations   32
Agent's Obedience   32
Agent's Care   33
Agent's Loyalty   33
Principal's Duties to the Agent   33
What Does an Agent Have Authority to Do?   34

**Differences in Legal Characterization: Independent Contractors
and Employees   36**

Business Organizations   41
Sole Proprietorship   41
General Partnership   42
Limited Partnership   45
Getting Out of or Ending a Partnership   47

Limited Liability Corporations   48
Corporations   50

Discussion Questions   51

CHAPTER

4

Legal Agreements: Contracts   53

Creating a Contract Through an Offer, Acceptance, and Binding Exchange   54

Proposing an Agreement: Offer   55
Accepting to Enter an Agreement: Acceptance   56
When Acceptance of an Agreement Becomes
    Effective   59
The Payoff: Consideration   61

Meeting Your Contract Commitments and Avoiding Common Contract Problems   62

Indefiniteness   62
Misunderstanding   63
Mistake   64
Unilaterial Mistake   65
Conditions in Contracts   65

When Courts Enforce Commitments Without Valid Contracts: Promissory Estoppel   66

Actions That Satisfy or Break Contract Commitments   67

Getting Out of Your Contract Commitments   67

Legal Exceptions for Not Completing Contract Duties:
    Impossibility, Impracticability,
    and Frustration   67
Miscellaneous Excuses for Breaking a Contract
    (Defenses to Breach of Contract)   69
    Illegality   70
    Duress   71
    Misrepresentation   71
    Unconscionability   71
    Adhesion Contracts   72
    Capacity   72
How to Cancel a Contract   73
Violating Contract Agreements   73

Treating Contracts Created from Unclear Bases: Oral Contracts, Statute of Frauds, and Parol Evidence   75

Statute of Frauds   75
Parol Evidence   75

What Might Result from a Lawsuit on Your Contract   77

Compensations to Winning Parties in Contract Suits   78

Courts' Limitations on Detriments Resulting
    from Broken Contracts: Foreseeability, Avoidable,
    Nominal, and Punitive Damages   79

Involving Others in Your Agreements: Assignment and
   Delegation of Duties   79
Third Party Beneficiaries   81
Ending Contract Responsibilities   81
Accord and Satisfaction and Substituted
   Agreement   81
Accounts Stated and Releases   82
Discussion Questions   82

CHAPTER 5

Intellectual Property: Trademarks, Trade Secrets, Patents, and Copyrights   85

Trademarks, Reputation, and Goodwill   87

Trade Secret   89

Patents   91

Copyright   92
Copyright Control   95
Legal Fiction of Employer Authorship:
   Work for Hire   97
Shared Authorship: Joint Work   100
Unauthorized Use of Copyrighted Materials:
   Infringement   102
Substantial Similarity   103
Defenses Against Infringement   104
Personal Use and Fair Use   106
New Developments in Intellectual Property
   in the Digital World   109
   Reproduction of Work in Digital Databases   109
   Deep Linking   109
   Digital Millennium Copyright Act   110
Discussion Questions   112

CHAPTER 6

Synthesis   113

Supportive Diagrammatic Schema   114

Case Context (Synthetic Situation)   117

Case Situation Analysis   118

Discussion Questions and Exercises   130

CHAPTER 7

Conclusion

131

The Complexity of Law and Ethics Applied 132

Case Situation 132

Case Context Questions 137

Works Cited 139

Index 143

LIST OF FIGURES

Figure 1.1   Benefits and Detriments of Varied Legal
              Characterizations   5

Figure 3.1   Debt Carrier's Reach to Sole Proprietor Assets   42

Figure 3.2   Debt Carrier's Reach to General Partnership Assets   45

Figure 3.3   Debt Carrier's Reach to Limited Partnership Assets   46

Figure 3.4   Debt Carrier's Reach to LLC and Corporation Assets   50

Figure 4.1   Elements of a Contract   60

Figure 5.1   Legal Protection for Intellectual Products   87

Figure 5.2   Work-for-Hire Elements   98

Figure 5.3   Constitutional Support for Intellectual Property
              Policy   107

Figure 6.1   Multiple Alternatives for Characterizing Intellectual
              Products   115

Figure 6.2   Additional Variable in Creator Characterization   116

Figure 6.3   Complications in Contracting with Creators to Develop
              Intellectual Products with Multiple Choices
              of Characterization   116

Figure 6.4   Agency Relationships, Control of Creative Products,
              and Parties' Legal Liabilities   117

Figure 6.5   MapTech's Interests, Liabilities, and Relationship
              to Spencer   127

Figure 6.6   Impact of Product Control on Liability   129

PREFACE

The working world has changed extensively over the last several years. Many workers no longer earn their livings within the confines of office spaces but telecommute from their homes. Many are as likely to do independent contract work as become long-term employees of structured companies. And many workers have become willing and even eager to create their own start-up companies to avoid traditional business structures. The image of the average American worker as the "company man" or woman toiling in a traditionally arranged office space has been replaced with multiple and varied images of workers in just as multiple and diverse work settings. These new forms of work arrangements and categories of new workers lead to a broader need among individual creators to understand the laws that control their work activities. And unique working arrangements can sometimes require new readings of currently applicable, but old, law. In particular, creative communicators who work in technical communication, graphic design, and digital development and communication, are likely to fall within the range of workers who need a simply structured basis for understanding how the law may affect them in their work arrangements. These workers are likely to need information for understanding the differences between law and ethics, contract law, the law of agency and business organizations, and intellectual property law. More importantly, they need a means to understand the law in these areas and how it directly affects them. This text was developed in response to these new needs.

What will creative communicators gain from this book that they might not gain from other books concerned with law and workplace issues? For one thing, there are very few books that treat issues in these areas as they apply directly to the needs of individuals who work in technical communication, graphic development, multimedia design, and other creative pursuits. Books that are not focused on these readers' needs but that could provide helpful information are often expensive, may not be comprehensive enough to cover the target audiences' needs, and are often written in legal language that can be difficult for lay readers to understand. For another thing, although these books explain the law, they often provide little material that will help readers understand how to apply the law to their own particular workplace circumstances. *A Legal Primer* is an attempt to overcome the obstacles of legal language, to provide a means for understanding and applying the law in its complexity, and to treat the areas of the law that are most specifically applicable to the audience of users noted above.

This book contains material most relevant to the workplace needs of creative communicators. Chapter 1 introduces the book and describes the content of each of the chapters in detail. Chapter 2 explains the differences in treatment of law and ethics and explains their different functions in the workplace of creative communicators. Chapter 3 provides extensive and comprehensive explanation of the law of agency and business organizations. It makes clear the rights and duties of people who work together and then describes the differing kinds of legal arrangements that workers can create to give structure to their working relationships (in the forms of partnerships, limited liability corporations [LLCs], and corporations). Chapter 4 explains contract law and describes the way that individuals can enter into legally binding relationships with each other, and also explains the potential effects of those relationships and the means for eliminating them, if desirable. Chapter 5 explains the differences between trademark, trade secret, copyright, and patent in intellectual property law, and notes how these different areas of the law are likely to affect creative communicators' intellectual products. Many books on legal issues provide case situations in which readers can apply the laws they learn, but Chapter 6 is unique to books treating legal issues in that it provides methods for understanding the applicable law within specific contextual situations. The chapter illustrates the law with desynthesized and resynthesized examples. The final chapter, Chapter 7, provides a series of questions that readers can ask and answer to analyze law in application. It includes a case situation to be used as a testing basis for information learned in previous chapters.

Although much of this book is informational in nature, the goal of the total structure is to provide a means for readers to apply in synthesis, within specific contextual situations, what they learned in the informational chapters. The last two chapters are devoted to explaining means for desynthesizing complex legal situations and then resynthesizing them with clear legal understanding. This approach is unique among books that treat legal issues and provides a new perspective for understanding the ramifications of the law as it is applied in context-by-context situations. The law can be explained in isolated information segments, but the potential outcomes can only be determined on a case-by-case basis. Each case, and its potential outcome, is shaped by the circumstances of the situation in which the legal context arises. Even one element that differs from case to case can affect the potential responses from the participants involved. In addition, each of the actors within a legal setting may choose among different actions that affect case outcomes. This book is designed to emphasize the complex contextual nature of law and provide a means for readers to treat it in application, thereby understanding this complex synthetic application of law to context.

Although *A Legal Primer* provides a basis for understanding the law and a structure for applying it in distinct contextual situations, it will not substitute for professional legal advice. This book is meant to lay a foundation for making legal issues less mysterious. It will give creative communicators a basis for gathering a general understanding of what to expect from applied law in their work settings. And although it is comprehensive from the standpoint of a lay reader, this book is not intended to provide material on all the intricate details of contract, agency, intellectual property, and business organizations law. Any one of the areas covered here requires extensive study and years of legal practice before applying the information in the actual practice of law. As such, this book is not intended to be a case study book for law students or an instruction manual for legal self-practice. But, these limitations notwithstanding, *A Legal Primer* should provide readers with a strong foundation for understanding the legal issues that will affect them in their workplace relationships and should help them to make informed choices that will benefit them far into the future.

Companion Website with Instructor's Manual

The **Companion Website** to accompany *A Legal Primer for the Digital Age* (http://www.ablongman.com/herrington), written by Jessica Cunard-Hunter, offers a wealth of resources for both students and instructors. Students can access detailed chapter summaries, a legal glossary, case studies (in contract law, intellectual property, and business relationships), and lists of relevant Web resources. Instructors can also download the **Instructor's Manual,** available exclusively in the website. The Instructor's Manual provides suggestions and guidelines for pedagogical approaches to *A Legal Primer for the Digital Age* and a discussion of the exercises and end-of-chapter materials. The Instructor's Manual will make instructors more comfortable with the complexity of the legal issues addressed in the book.

Acknowledgments

I heartily thank the many people who helped to produce this book. Particularly, I am grateful to have been able to work with Sam Dragga and Eben Ludlow, who invited me to produce this project and guided its development. I also thank William Russo, who has diligently managed all the difficult project details, and Michelle Campbell, who undertook the tedious job of copyediting, along with Shepherd Incorporated for composing this book. I especially thank Jessica Cunard-Hunter for her extensive work on the accompanying website. I am also appreciative of Laura J. Gurak's, Johndan Johnson-Eilola's, and James Porter's reviews and advice for shaping the final result. But especially, I am grateful to my parents, Jack and Pat Herrington, and to my patient husband, Yuri Tretyakov, who have unfailingly supported me in my efforts to complete the book project.

TyAnna K. Herrington
Georgia Institute of Technology

Introduction

The text you have begun to read was written primarily for technical communicators, multimedia developers, graphic designers, creative communicators, or students preparing to become communicators. But even if you do not consider yourself in one of these categories, you may find the material to be of benefit both in and outside of the workplace. The text is written in simple language and in a clear and concise form. It is also supported by a number of visual explanations and graphic examples. If you are in the target audience for the text, you will find that it covers issues in the legal areas that you encounter, or are most likely to encounter, in your everyday workplace activities, including contract, agency, business organization, and intellectual property law. Though these areas are covered thoroughly, the intention is to avoid the complexity of law that can be confusing and that can obscure the general principles that form the bases of legal reasoning. For that reason, the text is limited to a more general treatment of legal issues, rather than dealing with the intricacies of complex cases. If you encounter legal conflicts that require intense scrutiny, you are outside the range of purpose of this text and should seek individual help from a lawyer. But even then, the text will help you to better understand the broad concepts and principles that drive even the most complex of legal cases.

In your work as a technical communicator or other creative information developer, you encounter multiple elements of the law on an everyday basis. Consider, for example, a situation in which you are hired to analyze workplace communication efficiency and to provide the results of your report on an interactive website that allows space for user feedback. Your benefits and responsibilities would be determined by the legal relationship you have with the organization or person who hired you. If that organization or person is an employer, you would likely be legally bound to relinquish your copyright to the work you created, and—depending on the business arrangement within which your employment arises—you might or might not benefit from profits generated by the work or be legally liable for any harm caused by the work. Your agreements (contracts) with the hiring party would help to determine the business structure within which you would operate. That business relationship would further determine your rights and duties within the organization and the legal relationship to your employer, which would also affect your intellectual property rights. In addition, your ethical duties to the people and organizations with which you do business would be affected and also would affect all the potential relationships created by your contract arrangements. This example should make clear the importance of understanding contract, agency, business entity, and intellectual property law and its impact on the ethical questions that you address in your work.

In your experience as technical communicators, multimedia developers, graphic designers, creative communicators, or students preparing to become communicators, you may have encountered legal questions or problems that you found too complicated to understand. Your response to the law is not unlike that of many others, and rightly so. The law is a difficult, complex system for dispensing with conflict and providing order in society. It is based on a long history of British common law, which later was reiterated in American law through legislative statutory creation and case treatment. Because the law operates within contextual situations that are influenced by society's changing standards, there are no set right or wrong answers to questions of law. The same written law must be applied over and over in different contexts with different kinds of participants, whose life circumstances affect the way they act and fail to act. Courts must consider these varying conditions in their decisions. Most courts try to apply the law as consistently as possible, but each case has its own complications and intricacies, which courts must consider in order to make judgments. Therefore, the outcomes of cases vary considerably. Nevertheless, there are patterns in decisions, called "legal precedent," that help individuals who, often with help from lawyers, make their own best judgments about the likelihood of the legal outcomes of the actions they take. This text provides you with explanatory frameworks for making those judgments.

Even in best-case scenarios, legal situations are complex because the many factors that life provides come without labels for the categories of law that can be applied. For example, where you might think you have a conflict over an agreement, or contract, you might find that the other organization or person involved in the conflict treats the same issue by applying agency law. There are times when more than one kind of law applies and when a decision in one area can build to a potential decision in another. In other words, you can win a case on one ground, which may cause you to lose in another more detrimental way on another ground. With this in mind, this text provides not only linear explanations of areas of law but also includes synthetic treatment, in Chapter 6, of the legal issues presented in the book.

But the goal of this book is not to provide information that will help you win legal cases. Rather, it is to help you understand the law that is likely to affect you so that you can avoid legal conflict altogether. The more knowledgeable you are before you enter into legal relationships with others, and especially in the workplace, the more you will understand what you are agreeing to do, what your rights and duties are, and what to expect from those who work with you, for you, and who receive the benefit of your work. Understanding the consequences of your choices can go a long way in helping to make them.

Chapter 2 provides an explanation of the differences between law and ethics, and discusses four tests for determining whether behavior may be considered ethical. It provides a long series of scenarios for unethical behavior and then applies the four tests to analysis of case situations. Chapter 2 also supplies a case situation for your own analysis and asks a series of questions to lead you through the process.

Chapter 3 explains the law that is applied to business organizations and the relationships between and among people and organizations that work for, and with, each other. The chapter explains these relationships, called *agency,* and the business organizations within which the relationships operate by defining agency and by noting how to create an agency relationship. It then describes the different participants in agency and how they can be characterized for legal purposes, which indicates their rights and duties. To give you a basis for deciding whether you want to become a part of agency, the chapter also explains how legal liability is attributed in agency relationships. Chapter 3 also provides explanations of the differences between the legal characterizations "employee" and "independent contractor" and notes the effect of these characterizations on legal liability, as well as on control of intellectual products created within the relationships between and/or among hiring parties, employees, and independent contractors. The last part of the chapter describes and differentiates business entities in the forms of sole proprietorships, general partnerships, limited partnerships, limited liability corporations, and corporations. Chapter 3 provides a good starting point for deciding where you stand legally in your work relationships with others.

Chapter 4 provides you with explanations of the agreements, or contracts, that you might make, both orally and in writing, to help you understand how your contractual relationships affect you and others. The chapter starts by explaining the three elements required for a contract: offer, acceptance, and consideration. It examines how contracts can be broken in a way that is supported by law, how they can be broken in a way to cause harm to one of the parties, and how each of these circumstances can affect each party's ability to keep from bearing financial or other damage in the process. You should be able to use the material in this chapter to understand the potentially legally binding effects of your agreements with others, even though they may seem insignificant at the time of agreement.

Chapter 5 treats intellectual property law, which probably has the most direct and broadest effect on your work as technical communicators, multimedia developers, graphic designers, or other creative communicators. Central to your work is that you create intellectual products on a daily basis. Intellectual property law determines whether you, your employer, or another party has the right to control the products that you create. The chapter begins by explaining who retains rights to benefit from intellectual products and how the products can be characterized in trademarks, reputation and goodwill, trade secret, patents, and copyright. It also explains what effect their characterization has in law. The predominate focus of the chapter is on copyright, because this type of protection is most prevalent in treating the work you do. The chapter provides comprehensive information in this area, covering who controls copyrighted work, how joint works operate legally, how to determine cases of potential infringement, and defenses against infringement claims. The chapter ends with an explanation of fair use, an area of copyright law that may be important to you personally as well as in your capacity as a participant in commerce.

Chapter 6 provides a synthesis of all the information areas covered in previous chapters by presenting a lengthy case situation that demonstrates their complexity as they are intermingled in practice. The chapter includes a series of supportive explanatory diagrams to help you understand how to desynthesize the facts and issues in complex legal situations in order to understand their impact.

Chapter 7, the concluding chapter, includes examples and case studies that demonstrate the information provided throughout the

Benefits/ Detriments	As Independent Contractor	As Employee	As Sole Proprietor	As Partner	As LLC Manager As Protected Holder	As Corporation Manager As Shareholder
Control over intellectual property, product design, work schedule, employee choice, materials choice, work location	You control	Employer controls	You control	You share control	Managers, not holders, control	Managers, not shareholders control
Economic liability	You are liable.	Employer is liable.	You are liable; your personal assets are reachable.	You are liable for yours and your partners' part of debts; your personal assets are reachable.	Managers can be liable for negligence. Managers' and holders' personal assets are protected.	Managers can be liable for negligence, managers' and shareholders' personal assets are protected.

Figure 1.1 Benefits and detriments of varied legal characterizations

first five chapters. It points to the differences between law and ethics and notes some of the ethical issues that technical communicators must consider. It summarizes the work in a way that attempts to answer questions that might arise after you have read the earlier parts of the text.

Discussion Questions and Exercises

1. What is the law and what is its function in society?
2. What have been your own experiences with the law?
3. What areas of the law do you hear most about in the news? How do these areas relate to the fields of technical communication, multimedia, digital design, and other areas of creative communication?
4. What do you anticipate would be the most important area of law for you to become familiar with? Why?
5. To begin thinking of the potential legal issues that can arise in your current or future work in technical communication, multimedia design, digital development, or other creative communication, create a brainstorm document. Make a list of potential problems, and benefits, another list of work goals such as the kinds of projects you would like to develop or business arrangements you would like to create. Then list the legal and personal assets you will need to achieve these goals. You can use this document to help guide you through the rest of the text as you hone your understanding of law as it applies to your specific needs and interests.

A document that exemplifies the kind of problems/benefits list you might create is shown on the previous page.

CHAPTER 2

Law and Ethics

The primary function of law is to dispense with conflict or avoid potential conflict by following a set of rules and precedents that delineate expected behavior. The rules (laws) that make conflict resolution possible are developed through a negotiation process among members of Congress, whose participation is shaped by public influence. In many cases, laws provide a basis to prevent one party from taking advantage of another, but may be inadequate to maintain justice and instead may only dispense with conflict, the primary goal. Therefore, it is important to note that law and ethics are not the same thing and that legal behavior may not always be ethical behavior. *Black's Law Dictionary* defines ethics as "of or relating to moral action, conduct, motive or character, as ethical emotion; also, treating of moral feelings, duties or conduct; containing precepts of morality; moral. Professionally right or befitting; conforming to professional standards of conduct" (496).

There is no single test for determining whether behavior is ethical. Technical communicators, multimedia designers, and others in business settings will find that there are differing practices for treating ethics questions and that in some workplaces, discussion and guidelines for ethical practice are ignored altogether. Creators in the workplace respond in many varied ways to questions of ethicality, often relying on parental training or intuitive instincts to make decisions (Dragga "A Question of Ethics"). Among technical communicators, however, there are three common approaches for testing ethicality: universalization, common practice, and the utilitarian test for balancing the potential for harm against benefit, which weighs in favor of the greatest number of affected individuals. I suggest an additional approach, the axis-of-power test.

The ethical application of universalization is based in Immanuel Kant's test for whether an action, if universalized to a broad population, would have harmful consequences or maintain its sense of justifiability. He applies universalization in circumstances of "categorical imperative," the concept that some actions are absolutely right or wrong and that the categorical imperative demands that you choose the right/ethical action. His directive is that a categorical imperative is based on logical reason from which an absolute sense of ethicality must follow. The alternative, "hypothetical imperative," provides a means to use reason to determine that an action is ethical in accomplishing some specific end in one context, but its implication is that the particular action may not be ethical in other contexts and thus, not universalistic. In contrast, categorical imperatives are those that are universalistic and always ethical (Kant 70). To use the test, you would theorize what would happen if you repeated the act in question universally; if the detriment would apply universally, the act could not pass the test and could not be considered ethical.

The common practice or "existing practice" (Walzer) test tends to spread the denotation of "ethical" to a broader category of activity than the universalization test does. The common practice test excuses behavior as ethical if the behavior is commonly used within the field, workplace, or other organizational setting in which it occurs. And it allows communicators to "foster false inferences," or "produce false implicature," which lead people to assume conclusions as true when they are not. This practice allows communicators to mislead but not to lie (150), even though the effect can be the same. Those who consider this form of communication acceptable hold readers, viewers, or users responsible for arriving at false assumptions. They avoid placing any responsibility for misunderstanding on the communicators.

The concept of common practice is also applicable within a field of competing businesses. In businesses where technical communication managers are selling products and are expected to treat prospective clients to "percs" such as expensive dinners and entertainment, competitors would consider this practice ethical and would be expected to follow suit in order to be competitive in the same market. In business situations where this practice is considered unacceptable, particularly where it could lead to giving unfair advantage to one company over another, the common practice test would result in a finding of unethical behavior. When clients are political officials whose responsibility is to the public, they are limited in what they are allowed to accept as favors, and the practice of providing and accepting favors, would be unethical.

The utilitarian test of balancing weighs the benefit of an action against its detriment. If the benefit outweighs the detriment, the behavior is considered ethical so that you "measure [the benefit] for the appropriate number of people, compare it to measures of ill effects for the remaining people, plug it all into an algorithm, and calculate the solution" and end up with a "calculus of ethics" (Dombrowski 54). For example, if a technical communicator in a chemical company completed and printed five thousand copies of a brochure that included inaccurate information about the heating danger-point of a solvent, he or she could balance the potential harm against the potential benefit of the brochure. If the temperature was inaccurate by one degree, the likelihood of an accident caused by someone using the solvent at the incorrect temperature would be low, and the benefit of providing information telling the reader about the heating danger-point of the solvent would outweigh the detriment. In this case, providing the brochure, even with the inaccurate information, could be considered ethical. But where the balancing test often fails to provide a solid means for treating ethics in practical application is when it is used to test economic balance between benefit and detriment. In the scenario described, the technical communicator might use the test to balance the cost of reprinting five thousand brochures against the risk of the company paying for its legal responsibility (liability) as a result of someone being hurt by overheating the solvent. A decision to print the brochure based on this form of the test could be considered unethical. (See Markel for more detail.)

Note that the result of balances can be different for actions within and outside the workplace. For example, within some media companies, co-workers copy each others' code and product designs as part of their everyday work. Within these kinds of business settings, employers encourage a structure of sharing among employees to expedite product development. The common practice of code-sharing in this setting would be considered ethical, because it is part of the everyday process of doing business. In contrast, code-sharing outside the workplace could be considered ethically questionable if not a part of common practice. (Code sharing is becoming accepted practice among groups who participate, for example, in the open source movement, which aims to provide source code to all individuals freely so that they can enhance and recontribute new code to the user/coder community).

Although not cited as commonly used in the field of technical communication, I propose that you apply the "axis-of-power" test for determining ethical behavior as communicators. The axis-of-power test is based on the concept that those who have power to act or communicate information to others and are either obligated or choose to do so, also have responsibility to act and communicate honestly and completely without acting in a way that hides misdeeds or that misleads or masks information. In this way, communicators will not take advantage of others by misusing the power that comes with having information or controlling behavior (Herrington "Ethics and Graphic Design").

The axis-of-power test differs from the other three discussed. This test does not depend on a universalization of the effect of action or communication to others because it must be employed in context-specific circumstances. For instance, in a context in which a document recipient has as much information as the communicator does, the communicator may ethically represent that information in any way that he or she desires. The communicative action itself is not universalistic. So, in a different context, where the recipient has much less information than the communicator, the communicator is obligated to present the material in a very straightforward and clear way in order to meet ethical standards. He or she must adjust the nonuniversalistic communicative action in order to do so.

The axis-of-power test also differs from the test for utilitarianism balancing. The balance between benefit and detriment is never weighed by this test, although the balance of relative power between communicator and recipient is considered. Similarly, common practice is not considered even though some elements of the concept relate to the axis-of-power test. Whereas the common practice test bases its ethical outcomes on the assumption that if a practice is common within the action or communication setting of the participants, all will know what to expect and essentially be on even footing, the axis-of-power test is more specific. It provides that within each individual context, all participants in the shared setting must be on equal footing, even when that calls for the actor or communicator's greater openness in providing clear, honest information or undertaking honest action.

As a technical communicator, media designer, or other form of creative communicator, you will, by the nature of your work, have access to more information than your readers/viewers. This gives you the ability to control the information that you have by deciding how much and in what form you provide information to your readers. Your control over the information gives you a high level of power relative to that of your readers/viewers, and this power, in turn, gives you the ability to control them (within the context of the document that you provide). For example, if you are producing the annual report for shareholders of a corporation, you will know the details of the corporation's financial losses and gains, its failures and successes. You are responsible to the corporation for the work you do, and its board and managers will ask that you project its earnings and losses in the most favorable light. The shareholders will want enough information to determine whether their investments in the corporation will be likely to earn them money or cause them financial loss. As the communicator in control of corporation information, both favorable and detrimental, you have the power to control what readers/viewers understand about the company and their investments. If you mask information or mislead shareholders regarding the true value of their holdings or of the company they have invested in, then under the axis-of-power test, you have behaved unethically.

In law, as in every other area of life, the more knowledge a person has, the greater his or her advantage over someone with less knowledge. This is particularly true when there is an imbalance in understanding the laws that govern business arrangements. You should be aware that it is possible to misuse the power that comes from understanding the operation of laws to put another at a financial or other kind of disadvantage. Lawyers, with their special training in the law, have explicit duties to benefit their clients and to refrain from using their special knowledge to overpower those with less understanding of the law. Lawyers are bound by laws based on the American Bar Association's and other associations' codes of ethics. They are also bound by federal and state laws that define attorney misconduct and allow judges to discipline unethical conduct.

It is also possible to cause harm in a manner that is legal but unethical. Laws can allow action such as broken promises, disloyalty, and many forms of advantage-taking that are not illegal but that have the potential to cause great harm. And laws exist that, in themselves, are grossly immoral and ethically unsupported, such as the Fugitive Slave Act of 1850, which required Americans to return runaway slaves to their masters, and the U.S. Supreme Court's Dred Scott decision, which declared that slaves were not citizens but property (60 US 393 (1856)).

As a technical communicator or multimedia designer, you may face choices that could lead you to make decisions that are legal but unethical. For instance, you may find that you have difficulty squaring your sense of personal ethics with your sense of loyalty to your employer or to the workplace within which your job status and image is developed (Clark and Doheny-Farina 470). You may also encounter situations in which you face unethical attempts by others to take advantage of you. Among the many possibilities for legal unethical behavior are the following:

- In some situations, attorneys are claiming that use of regular Internet E-mail is a waiver of the attorney-client privilege, because the Internet is not a secure medium for private conversation (Kuester). Although this argument may avoid a violation of attorney-client privilege, to divulge a client's private conversation with his or her attorney when the client expected the conversation to be protected could put a client into a situation that he or she entered involuntarily. Where open conversation with an attorney can be induced by the belief that the conversation is private, the attorney has power over the client, coupled with a duty to protect the client. An argument that use of E-mail negated the privilege might hold up under legal scrutiny but certainly not under ethical analysis. Technical communicators and multimedia designers use E-mail as a communication tool with as much or more frequency than they use the telephone or face-to-face meetings. The likelihood that you might unwittingly fall into the circumstances described above is higher than it would be for those whose E-mail use is not as prevalent. As such, you should be aware of the potential for abuse and guard against the possibility of falling victim to unethical behavior.
- In all but six states (Illinois, Massachusetts, New Jersey, Rhode Island, South Carolina, and Wisconsin), lawyers who have been disbarred for ethical or other violations are legally allowed to practice as paralegals. Where it might be tempting to save money on lawyers' fees and hire a paralegal to work for you instead, you might be well served by checking the backgrounds of the paralegals you work with in order to avoid working with one who may have been involved in unethical or illegal behavior in the past.
- Individuals well versed in the law know that it can be very difficult to prove certain types of legal claims, such as those for sexual harassment. There are also a number of claims that may be so costly to pursue that spending the time and money to pursue them would cost more than the benefit of winning a suit. When a person uses this kind of knowledge to avoid legal sanction, he or she is not only participating in potentially illegal behavior but also in unethical behavior. You should be aware of the possibility that you may encounter this kind of unethical behavior and that it may involve an employer asking or ordering you to participate.
- Whether you are or will be a technical communication or multimedia business owner or employee, you should be aware

that it is not illegal for your business's Internet service provider (ISP) to monitor the content of your E-mail conversations. In fact, under the Digital Millennium Copyright Act (DMCA), an ISP is liable for the effect of content it controls, and thus, is implicitly required by law to monitor content.

■ As you will learn from Chapter 5, the law allows a copyright holder to control the information that he or she has created. The holder may license use of the copyrighted material under his or her discretion and may also limit the extent and purpose of use. However, limiting access to copyrighted material to avoid critical comment of it or the creator is questionable both legally and ethically. In the last few years, the Church of Scientology has been cited many times for attempting to use copyright and other laws to prevent or attack critical evaluations of its teachings and practices, and to avoid legal efforts to create guidelines for governing what has been seen as its unethical recruitment tactics and fraudulent fund raising, among other potentially illegal or unethical actions. (http://www.gospelcom.net/apologeticsindex/r04.html)

■ You may someday be employed in a company that works with dangerous materials. Materials use and waste disposal are regulated by local and federal laws. In some instances, the technical letter of the law regulates chemicals by specific name, but companies may also use newly developed chemicals not on the list. Technical legality may not require the same level of care with their treatment and disposal, but ethicality does. You may be asked to write about materials that reflect this disparity in the treatment of dangerous chemicals and will have to make decisions based on your own assessment of ethicality.

■ You may also be confronted with the prospect of overselling a project to a client to get a job—promising expectations that you may or may not be able to meet or taking advantage of the lack of specific contract details to shirk client expectations later. Unless you actually breach a contract with terms that note expectations, you will not be acting illegally, but you certainly will be acting unethically. Moreover, it would be to your and your company's disadvantage to promise results that you cannot produce. Good business practices depend on reputation, and every relationship with a client adds to the reputation that you build within your work community.

■ You will likely encounter a telemarketer at some point in your career. Telemarketers try to entice you to buy products (usually office supplies) that you would not buy otherwise or to buy products from them rather than from your usual supplier. Their pitches may not be illegal but can be highly unethical. Telemarketers often gain an advantage when they talk with an inexperienced employee who is unaware of or does not completely understand the employer's standard purchasing procedures. Thus, if you own your own company or if you are responsible for office workers, it would be helpful to train them to avoid telemarketers before they respond to office calls. Telemarketers may use an authorized purchasing agent's name or the name of an employee well-known within the company to gain the trust of employees who order supplies and other materials. Some may only ask for an order, where others may entice an employee to divulge information that should remain within the company. Telemarketers who go as far as to misrepresent themselves as working for another company, such as your regular supplier, cross the line into illegality by acting fraudulently. Most often, they use their own company name, often one that sounds impressive or sounds much like that of one of the big suppliers. The harm in responding to telemarketers is that they may provide inferior products or sell you more supplies than you need or can use.

■ As you will learn from Chapter 3 on agency, an employee (agent) has a legal duty to his or her employer to act for the employer's benefit. But there are times when disgruntled employees act to harm the employer, and do so in such minor ways that there is little basis on which to claim illegality. Their actions are, nevertheless, unethical. For instance, technical communicators make choices in design and layout when developing documents within the range of their special expertise. They choose typography, layout, graphic inclusions, and color schemes for print and web designs. Because they are trained in this area, employers rely on their judgment to create documents. In the instance that a technical communicator has no particular ink color preference for a print document but chooses red, a relatively expensive ink, to retaliate against an employer, the technical communicator is acting unethically.

■ In an age where copying code and layout choices from the Web is a common and well-accepted practice among web and other multimedia designers, the ethicality of copying code and design features is difficult to determine. As you will learn from Chapter 5 on intellectual property, computer code is treated like language today and not generally copyrightable, so it is not legally protected from copying and displaying. But a scheme of code can represent a company and using another company's design can be unethical. In addition, web designers are hired for their special skills and unique artistic talents. If they copy code rather than acting from the basis on which they are hired, they fall short of their obligations to the employer. They also produce nothing of unique value upon which to build their own careers.

- Workplace settings can be hectic, stressful, and very demanding. When technical communicators and multimedia designers become overwhelmed with work, they sometimes become desperate to find a way to catch up with demands. New employees often experience pressures both from the usual demands of the job and the adjustment to a new setting, and make an extra effort to please their co-workers. When seasoned employees pressure new employees to take on more than their share of work, their actions—although not illegal—may not be ethical. Not only is this an abuse of power, but the action may end in the loss of a potentially helpful and able new employee to a less stressful position, which hurts the company in turn.

- New technologies bring new kinds of business competition as well as new means to access potential customers and clients. Although in most cases it is not illegal to sell client telephone and E-mail address lists to companies that wish to solicit business, this practice is potentially unethical. Very few customers claim to enjoy telephone solicitation calls or E-mail "spams." Where customers either want or agree to tolerate solicitation, selling their information is not harmful, but making customer contact information public when it is not desired, even though customers did not directly express their desire to avoid solicitors, is unethical. In addition, clients and customers who learn that their contact information is sold without their permission may respond by dropping their relationship with the company that sold it.

- Sales of space for web site pop-up ads may also lead to loss of customers and can be questionably ethical. When users browse free websites, they have little to no expectation to maintain uninterrupted access free of advertising. But when users pay for site use, either for access to information or for a service function, they expect to do business without outside solicitation. If you are in, or are entering, a career as a web designer, you would be wise to consider your design choices carefully, with this question of ethics in mind.

- Web designers must also take care not to design pop-up boxes that users inadvertently click on without knowing that they have agreed to receive "spam" or to be drawn into some other form of solicitation. Designers are responsible to their users to make clear what users agree to when they enter the site. Web designers must remember that they, and not users, know the content of a site, the intended outcomes of its use, and the results of every user action. Thus, web designers have both the power and the responsibility to provide clear, well-defined content.

Although the material in the following chapters is only a preliminary introduction to contracts, agency and business, and intellectual property law, you will likely gain a broader knowledge of legal issues. A greater understanding of the law will provide you with a relative advantage in dealing with problems that you may encounter in the legal areas discussed. This book's goal is to help you understand legal issues so that you can better protect your interests, plan for your future, and become comfortable with the business organizations that you will or do work within. But as a result of your greater knowledge, you will also carry a responsibility to consider the ethical effects of the choices you make in using that knowledge. In other words, you also will have an ethical duty not to use your advanced knowledge to take advantage of others.

Legal procedures are extremely expensive, primarily due to costs for lawyers' "billable hours," the monetary notation for the time they spend working on cases. In standard pay scales, fees increase with each step clients take toward litigation, and lawyers' fees are sometimes double the initial amount if a case goes so far to enter court proceedings. Other expenses in areas such as information gathering, witness testimonies, and filing fees can add to the financial burden of legal proceedings, but the bulk of expense resides in the work of the lawyer. Because even defending a suit can be a huge expense, there is a risk that unscrupulous clients and their lawyers might file claims as a tactic meant to scare defendants into acquiescence to client demands. To combat the possibility of abuse, Rule 11 of the Federal Rules of Civil Procedure (28 U.S.C.A.) is intended to protect defendants by punishing lawyers and clients who file frivolous or abusive claims in court. Nevertheless, the fear of potential legal trouble can be enough to inhibit legal, rightful, and sometimes protective behavior from individuals who balk at any claim of potential illegal behavior. Therefore it is helpful to gain as much knowledge as possible about business activities before you enter into them and once you gain enough knowledge to protect your own interests, maintain your responsibility not to abuse that knowledge to harm others. The following are examples of potentially unethical behavior stemming from knowledge of the law:

- It is not uncommon to find statements claiming broad intellectual property protection for materials posted on web sites or in print. As you will learn from reading Chapter 5, the fair use doctrine protects users' rights to access copyrighted and otherwise protected materials. A claim that it is illegal to copy material posted on a web site for any purpose is not only misleading but is wrong. The effect of these kinds of claims can be to scare users from exercising their rights to access (often called an "in terrorem" effect) that are granted by the copyright law itself. Where a claimant misuses the law to make broad claims of protection, the result can be to hamper users' rights and inhibit free speech. You should be aware of, and not participate in, unethical behavior resulting from making false claims about the power of the law.
- By the same token, users can also act unethically by copying and distributing creators' works outside the bounds of what fair use and the first amendment allow. Users who know that it can be difficult to prove claims of intellectual property violation may go beyond the bounds of what the law allows, resulting in unethical behavior.
- In addition, where users copy others' materials and claim them as their own, they engage in plagiarism and further violate ethical bounds.
- You may encounter a company—or create your own—that keeps an attorney or law firm on retainer. To work on retainer means that an attorney or firm maintains a continuous association with the company that retained it and is usually paid a set fee on a regular schedule, regardless of the number of legal issues that arise. Although most attorneys or firms make arrangements for additional fees per legal dealing, where a company has a high volume of legal dealings, the money that it spends to handle its legal problems is appreciably less than it would be without the retainer relationship. Organizations that use a retainer relationship to pursue suits against others as a means of coercion, knowing that the other parties' financial power to defend themselves is limited, act unethically. Financial advantage is not unethical, but its misuse to bully and coerce others—particularly those who may have valid legal claims—is unethical. For example, a company hires a contributor to develop what the contributor assumes is a joint work, where she would have an equal interest in the product. The

work is of minimal value but has the potential to be valuable in the future. Knowing that, company managers ask the contributor to relinquish her right to the work as a joint work. When she refuses, the company, using the power of its retained attorneys, claims that the work was not a joint work but a work for hire, and that the contributor has no claim. Because the product has only potential value and because hiring an attorney would be extremely expensive for the contributor, she is unethically coerced into relinquishing her claim to a joint work.

▪ Agents may act unethically by using a company name, not to harm the company or gain financially in the company's name, but for their own purposes, such as to gain access to financial or professional opportunities that they would not have otherwise. For instance, a multimedia designer might gain free entrance to an invitation-only event at the annual SIGRAPH conference on digital design by using a listed company name. If the employee was asked to participate as a company representative, his or her entrance would be expected and, of course, ethical. But if the employee used the name to gain entrance without the company's sanction, the behavior would be unethical.

▪ On the other hand, an employer may use an unknowing agent to carry out questionable activity. Employers often know better than their agents the gray areas where illegality is difficult to determine. Uneducated agents may rely on their employers and pursue behavior that is questionable without knowing the potential impact. Ultimately, as you will learn from Chapter 3, employers are legally responsible for the acts of their agents, but agents still must bear the burden of a harmed reputation, even if they are not declared legally responsible for illegal acts.

▪ This chapter demonstrates that law and ethics are not the same and do not operate in the same way. This concept may be difficult for some people to understand and may lead them to believe that as long as they follow the law, they have no other obligation, ethical or otherwise. However, acting within the range of legal technicality while pursing activities that potentially may harm others, although legal, is nevertheless unethical. For instance, a group of multimedia designers, technical communicators, and graphics designers might join together to create a corporation for the purposes of using word-of-mouth hype to sell stock at inflated values and then sell the stock while the value is high. As long as they did not misrepresent their claims about the company, its holdings, and activities, their actions would not be illegal (http://www.capitalism.org/faq/stocks.htm). But their motives to induce a false stock valuation and their taking advantage of individuals' lack of knowledge of real motives for creating the corporation would be highly unethical.

- Some may act both illegally and unethically at the same time. There are shrewd and often knowledgeable individuals who take advantage of the complexity and expense of legal proceedings to participate in wrongdoing, knowing that it would be more expensive for a claimant to take legal action than to ignore the illegal action. For example, a technical communicator might create a corporation's annual report and include visually misleading and fraudulent information that induces readers to engage in financially harmful activities. Where the visual material may have misled readers, the textual materials nevertheless represented the truth about the company, making it very difficult and expensive to prove that the technical communicator created a false report, even though the communicator's intention may have created fraud.

You may question the possibility that you can control your own approaches to ethics within workplace settings where behavior is demanded or expected that you may not find ethical. Or you may find yourself in difficult situations when your ethical choices seem unclear, and you feel that you have no basis for establishing accepted behaviors within your workplace. But as communicators, their employers, and co-workers have begun to consider the humanistic qualities of their job sites, they have also begun to develop a greater sense of authority for affecting their activities at work and have become aware of the responsibilities that come with it (Faber, Savage). Now that you have a sense of what kinds of ethical questions may arise and how to identify them, test your understanding by examining some areas of potential ethical problems. Remember that I previously described four approaches to deciding ethical issues: universalization, common practice, the utilitarian balancing test, and the axis-of-power test. To demonstrate how they operate, I will analyze the first contextual situation in four ways by applying the four tests listed here. I ask you to apply your own knowledge about ethics and analyze the other contextual situations that I provide in the same way.

Ethics Contextual Situations

In the first contextual situation, you are a technical communicator working in a small engineering firm whose work is to test small mechanical parts for safety. The company has been growing slowly over the years but from time to time faces financial difficulty. This month one of the company's most expensive testing mechanisms broke and not only did the work (and the fees for testing) stop during the time it was being repaired, the repair itself was very expensive. Part of your job is to write and submit the financial reports on the business. You work with the accountants, as well as others, to gather the information you need. Investors study the reports you provide and use them to determine their level of financial interest in the company. This month the accountant tells you that he manipulated the figures for the repair so that the information for the report would not reflect financial loss. He tells you that he will add the amount of the loss back into the reports by noting loss in small increments each month, even though the loss occurred all in one month. You hesitate and tell the accountant that you are not comfortable including inaccurate information in the report, but he tells you that his allocation of loss over several months is typical and that every accountant in business learns to do the same thing. He tells you that nobody will even notice the accounting ploy. You negotiate with the accountant, who provides you with an explanation about how he derived the accounting figures. You can then include this information in the report, but documented in such a way that it is visually inaccessible and will likely go unnoticed.

Remember that the universalization test speculates on the outcome of the unethical action tested if everyone were to follow the same practice. If the practice were to end in some kind of harm, the action would be unethical. In this case, if every accountant in every business reported financial loss over several months rather than the month in which it occurred, investors would have no way to know the values of the companies they invest in. They could put themselves at financial risk by investing in companies without solid financial grounding. Ultimately all investors could lose money, which would weaken the financial market, making further investment untenable. Under the universalization test, the harm would be extreme, so the activity would be considered unethical.

The common practice test determines ethicality based on identifying the normal, regular actions within the given context and on denoting as ethical those actions that follow the norm. In this situation the accountant has told you that his action in allocating loss over several months' time reflects what every other accountant does. If it is truly common practice in the business, then you would have to find the action ethical.

The utilitarian balancing test is more complicated. The test balances the potential harm caused by pursuing the questionable action against the benefit you will receive. If the benefit outweighs the harm, then the action is considered ethical. In a straightforward situation, you might find that it is ethical to decrease all employee salaries by ten percent in order to avoid a company failure and thereby allowing you to save the jobs of all employees. The harm to employees is minimal compared to the potential harm of losing their jobs altogether. But where the balancing test is used to balance financial risk against benefit, or potential economic loss against economic benefit, the results can be difficult to determine. In the situation described, if you balance the risk of the accounting mistake becoming public and decreasing the company's goodwill against the harm caused to the company if someone found out, the test might allow you to take the action. But then the test is not really one of ethics but a question of risk and probability. In a true test of ethics, you would have to balance the potential harm to investors who would not know of the loss against the potential benefit to them. Applying the test in this direction shows that ignorance of the loss is of no benefit to investors, so the action would be unethical.

The fourth test, the axis-of-power test, focuses responsibility for communicating information on those who have the power to choose whether to communicate it or not. In this contextual situation, you have the power to provide the truth to investors or to hide it. Technically, you will be able to provide the truth about the accounting method, and thus the loss. Functionally, however, readers will not be likely to see the loss reported because you will have misled them and masked the information. Under the axis-of-power test, you will have acted unethically by using your power to mask and mislead readers, even though technically, you will have provided the truth.

Keeping the operation of these tests in mind, how would each apply to the following contextual situations?

In your small multimedia business, you use the Internet for every aspect of your business. You and your partners spend so much time at work that you use the Web for playing games, personal communication, online shopping for personal items, and even for ordering pizza. You find that the expenses you owe to your Internet Service Provider (ISP) are extremely high, and you need to find a way to decrease them in order to keep your business on solid financial footing. You decide that because your Internet access is a part of your business expenses, you should be able to pass off the charges to your clients. In every new client contract, you include an expense line indicating that the amount noted pays for online fees. Would this action be ethical under each of the four tests?

In your position as a technical communicator in an engineering firm, you have written a joint work, a set of biographies that will be included in an insignificant book on the history of the company. Your supervisor had asked you to research the chief executive officer's (CEO's) family history and include all the detail you can find. The CEO of the eighth generation of the family-run business asked for the biographies as a part of a book celebrating the firm's history. The book's real value is primarily personal to the CEO. In your diligent work on family history you included a paragraph about the CEO's illegitimate daughter, who has a legal claim in the business. Neither you nor your supervisor were given instructions to leave out the information, that was supposedly hidden, and your supervisor sent the material out for publication. The book was self-published and copies distributed to family members and one hundred employees. When the CEO found that the history of her illegitimate daughter was uncovered and published, she was livid and commanded that all copies be collected and returned to her. But a business competitor had already been given a copy and learned about the former secret. Now, the competitor asks for copyright license to reprint this part of the book because the story has leaked out, the illegitimate daughter has reappeared, the story has become a community fascination, and the competitor is writing a new work on the history. Your CEO refuses but licenses the same material to a television company that plans to create a made-for-TV movie that is extremely flattering and sympathetic to the CEO. What are the ethical issues involved here and how would they be treated under the tests?

You are a management supervisor of a writing team in a small corporation and have just learned that your company will be merged with another larger one. The information was given in confidentiality to help you prepare to weed out your writing team to accommodate the inclusion of more seasoned professionals from the merging company. Leaking information about the merger could affect the stock values of both companies and inhibit the merger as well. Your close friend, a good writer and a hard worker who is also new on your team, has told you at lunch about her child's recent diagnosis of long-term illness and that she is considering taking another job in a company that provides a higher contribution on insurance benefits. She tells you that she has thought long and hard about the decision, but out of loyalty to you and the company has decided that on the following day she will turn down the job with the other company. What do you do?

As a multimedia designer, you have a new client—an old company that wants to shed its stodgy image and appeal to younger, more trendy customers. You have been asked to use photos of company representatives that were taken 20 years earlier and to update them to look like new photos in order to give the impression that company representatives are younger than they actually are. What would each of the four tests lead you to do? Can you point out all the ethical issues in the case situation described?

Discussion Questions

1. Are there times when breaching a contract can be beneficial and still ethical? What would make breach of contract unethical?
2. Is it unethical to enter into a contract to perform work that you are not sure you are skillful enough to perform? What if you plan to learn the required skills before you perform your contract duties?
3. Could ethical problems arise from keeping a copyright secret? Could legal problems arise from the same secret? What would they be?
4. What are the fundamental differences between legal and ethical issues?
5. Which of the four approaches to ethics described in this chapter most closely approximates the way you decide ethical issues?
6. Which of the four approaches to ethics described in this chapter provides the most consistently fair results?

CHAPTER 3

Business Relationships and Organizational Structures

 In this chapter, you will learn about the legal ramifications of the relationships between and among the people you work with and for, your relative responsibilities and duties to them, and theirs to you. The legal characterizations of these relationships is called agency law. This chapter will first help you to classify your own role as a creator and to determine which legal characterization of your work and yourself will be to your best advantage. Then it will help you understand the potential business structures that you might enter into within the legal characterizations that you choose and will help you determine your most advantageous business relationship. Your understanding of agency law will help you note how important it is that technical communicators, multimedia developers, and other creative communicators understand the distinction between the legal characterization of "employee" or "independent contractor." Chapter 3 explains the 13 elements that lead courts to make their definitions of these legal characterizations and also explains the effects of the characterizations. It is important that you see that different rights and duties result from the differing characterizations as employee or independent contractor. Your legal characterization can determine whether you or the person or business you work for will be responsible for certain legal responsibilities, and it also determines what your other responsibilities and duties to your business or to individuals will be. In addition, your rights to control the creative products that you develop will be different if you are an employee rather than an independent contractor, and you should know how the law dictates your interests in this area of work as well. Included here, also, are descriptions of the different kinds of business relationships you might enter into in your work, as well as explanations of the potential legal effect of your participation in the different business entities. As a creator who will either work within another's business, provide freelance work for another, or create your own business as a basis for development of your work, you will find it important to know that the various types of businesses or individuals whom you work for or with will affect you in as many varied ways as there are business organizations and relationships. This chapter explains those businesses and relationships and the impact of the laws that govern how they and you must operate together.

Your Characterization in Business Relationships and Its Legal Effects

Agency Relationships

Technical communicators, multimedia developers, public relations and advertising creators, and others whose work is based in developing documents that communicate, all work either as part of a business structure or with a business structure or entity. You may already be in one of these positions, or you may be soon where you work alone, with other individuals or groups, or in organizations of workers. Whatever your business relationship arrangements, they create legal consequences. As long as people who work together meet mutual expectations and avoid conflict, their legal relationships and characterizations often seem invisible. But when conflict arises, establishing the exact nature of business relationships becomes important for determining the conflict outcome. And in light of the different legal characterizations of workers as employees and employers, independent contractors, and hiring parties, deciding conflicts can be difficult and complex. So it is far better that, before entering into a workplace relationship, you and other communication creators know your responsibilities, duties, and the legal consequences of doing so.

This chapter explains the fundamental aspects of business relationships and defines and describes the kinds of business organizations that communication creators may enter into both to protect their interests and to establish working relationships. Reading this chapter is by no means a substitute for working with a lawyer to create legally sound business organizations or to defend or pursue litigation once conflict between business entities occurs. Neither is the intent to provide every detail of the law of agency, partnership, and business organizations. Instead, but it will give you a road map to help you understand the general workings of business relationships so that when you do enter into agreements and business arrangements, you will have a better idea of the potential impact of your choices.

Your relationships with your co-workers and the business organizations that you work in and with often create what is described by the legal term *agency* (*Cox v. Hickman, Meehan v. Valentine*). In fact, agency is one of the elemental relationships that exist in workplace settings. When you agree to work for another person you create an agency relationship because you allow your employer to direct your actions or activities within the employment. Your actions as an employee both reflect on the employer and can bind the employer legally. In essence you stand in the employer's stead as an employee when you act under the employer's direction; thus, you are his or her agent. Agency must be voluntary and by consent of the participants, thus creating a relationship between two people or participating business organizations. Agency is supported by law whenever one person has a right to control the conduct of another. The party who controls is called the "principal" and the controlled party is called the "agent." The agent has the power to create legal relationships and obligations for the principal. Most often, agency relationships exist in workplace settings where one person hires another to perform some kind of service or create a product. Agents may be employers who operate businesses and hire independent contractors, consultants, employees, and window washers or cleaning crews—again, as agents. They also may be individuals who hire others to mow their yards or buy stocks in their names. An agency relationship can be created without a business organization as its base. For instance, a web designer may provide an HTML version of a resume for a friend who is unfamiliar with the Web and its operation. The friend may ask her to upload the file to his web site to make the resume accessible to potential employers. When she undertakes the work on the resume and uploads it to the site, she becomes her friend's agent. If it turns out that her access to the web space for this purpose was unauthorized, her friend is responsible for her actions. (See Restatement 2d of Agency §§ 1(1), 1(2), 2(1).)

Agency often becomes complicated when the agent, usually an employee, is authorized to perform a task that can affect the principal's relationship to another party. Legal conflicts commonly arise when an agent, in the name of the principal, has acted in a way that does damage to a third party, but when the principal claims that the agent acted outside the realm of what the principal authorized. Conflicts also occur when agents fail to carry out the duty of their agency or violate the trust of their principals and act in a way that causes the principals harm. For example, where a graphic designer, the principal, hired an employee, the agent, to deliver completed projects to clients on time and intact, and where an agent delivers materials late, smudged, dirty, or incomplete, this failure to meet professional standards reflects on the principal. In fact, if the principal had contracted with the graphics' recipient to deliver the materials on a specific date so that the recipient could meet her own contract obligation, the agent's failure to deliver them on time could make the principal legally responsible for the recipient's business loss resulting from his own late delivery of the materials.

The most important aspect of the agency relationship is that the principal is legally responsible for the agent's actions, within the authority of his or her agency, so that the agent's actions are attributed to the principal as if he or she had personally done the acts. This is often called "vicarious liability" or *respondeat superior*. But the principal is only legally responsible for the agent's actions if they fall within the realm of duties (the scope of employment) that the agent undertakes for the principal. The agent must also be the principal's employee rather than an independent contractor. An employee is treated differently from an independent contractor under the law, and the differences lead to consequences for both the agent and principal. The distinctions are discussed in more detail later in the chapter, but first you need to have a clear understanding of the construction of agency before considering distinctions. (Refer to the section regarding the 13-element test that determines employee status on pages 38–40 in this chapter for agent-independent contractor distinctions.) It is important to note that for legal purposes, the terms *employer* and *employee* imply specific legal consequences and must be distinguished from other terms denoting working relationships. Therefore, the terms *agent* and *principal* are used in this text, not to complicate your understanding of the material, but because, when discussing legal qualities, there is a significant difference between employee and agent.

How to Create an Agency Relationship

How can you create an agency relationship? Agency requires no formal documentation or formalization through a written agreement, although it is often wise to lay out the terms of agency in textual form so that parties to the agency are completely aware of their rights and responsibilities. Whenever, individuals develop an agency relationship for business purposes, they should consider textual documentation necessary. An agency relationship includes a principal and an agent. The principal controls the actions of the agent performed on behalf of the principal, and the agent has a duty to accomplish those acts on the principal's behalf. The agent is most often economically compensated for performing the duties of an agent, but may be compensated by other means. Principals and agents come in different forms. You will need to be aware of what each form is and how it operates so you can make choices in creating your agency relationships.

Agency-Principal Relationships and Mutual Duties

Three Kinds of Principals

Remember that principals are the people or business organizations that control the agent's actions; principals are most often employers. There are three kinds of principals: disclosed, partially disclosed, and undisclosed (Restatement 2d of Agency, §§ 4(1), 4(2), 4(3)). A principal is disclosed if an agent acts for a principal to do business with a third party and the third party knows that the agent is acting, not for himself or herself, but for his or her principal, and knows the identity of the principal. In other words, the principal is disclosed if it is obvious that one person or business is acting on behalf of another and the third person outside the relationship between the principal and agent knows the principal's identity. A principal is only partially disclosed if the third party knows that the agent is acting for a principal but does not know the principal's identity. The most secret arrangement of the three, the undisclosed principal, exists when the third party has no knowledge that the agent is acting for someone else. In other words, the third party assumes that the person or business organization he or she is dealing with is acting for himself or herself.

Different Kinds of Agents

There are also different types of agents in agency relationships. These include general agents, special agents, and subagents. A general agent carries out the broadest range of activity done for a principal. A general agent acts for the principal on a continuous basis and conducts a series of ongoing transactions for the principal. For instance, a web master may do all the ongoing web site maintenance work for a firm from month to month over a set term, or may do the work on a permanent basis over a number of years. In both cases, the web master may be considered a general agent. A general agent may also carry out a number of different kinds of duties for the principal. A multimedia department manager, as a general agent, may order supplies, software, and hardware, may study hardware specifications to make choices of what to buy, and also may hire employees and oversee their work.

The special agent's duties are more limited, often to a specific one-time task or to a specific type of task. A special agent carries out a single or limited number of transactions for a principal. A special agent may be hired to complete the principal's tax accounting for a given period. Or a principal may employ a special agent to buy three adjoining properties in a given location in preparation for building a new structure. A special agent's duties are limited to one kind of action, and the agency relationship is often limited in duration. If you someday own your own technical communication documentation firm, you will likely hire a tax accountant and will need to be aware that a special agent relationship is possible. You might hire the accountant to complete your tax returns for one tax period or for all, but the accountant's duties would be limited either way; thus, you would create a special agency.

In addition to general and special agents, a principal may hire a servant, another form of agent, to act on the principal's behalf (*United States v. Ronitti*). The principal generally exercises more control over a servant than over other types of agents. The principal can control a servant's physical conduct during the time the servant is employed, and the servant agrees to devote time to the principal's business and often to personal affairs. A principal's right to control this type of agent's activities creates a "master-servant" relationship, where the master is legally responsible for the results of the servant's activities, as long as the servant acts in the service of the principal. This legal condition of nearly absolute legal responsibility justifies the principal's power over the servant's activities. For instance, the manager of a prosperous technical communication company who hires a personal assistant may create a master-servant relationship if the manager and servant agree that the servant's physical conduct is placed under the manager's control. Therefore, the servant drives recklessly when running an errand to pick up the manager's laundry, the manager (master) can be held legally responsible for the servant's reckless driving.

Despite the need for a principal to show control of the hired party to create the legal structure of agency (*Ahn v. Rooney, Pace, Inc.*), today there have been moves to create an "electric agent." Although not currently in force, the latest draft of the Uniform Commercial Code, Article 2 of the American Law Institute's Discussion

Draft, states that an "electric agent" is a "computer program designed to initiate or respond to electronic messages without review by an individual." This kind of development resulting from digital communication is an indication of what may come in the future in relation to agency arrangements.

In contrast, an independent contractor may conduct transactions for a principal but not be subject to the principal's control. An independent contractor is thus distinguished in an important way from an employee (Reuschlein 16). In addition, the principal is not liable for the independent contractor's actions. Independent contractors use their own judgment to meet the responsibilities of their duties to the principal and most often have special knowledge that makes them capable of accomplishing what the principal could not accomplish alone. For example, a digital artist might be hired as an independent contractor to create characters to be used in a principal's advertising campaign. The principal would not direct the artist's choices of software or hardware, and would not control employment of additional digital artists to help. In most cases, independent contractors supply their own equipment and supplies, work from their own places of business, and dictate their own working hours. Independent contractors, as their title implies, work independently from the person or business organization that hired them and are treated independently regarding their actions under the law.

Subagent relationships exist when one agent hires another to complete the principal's business. An agent can create a subagency only when he or she is authorized by the principal to do so. The principal is not held responsible for the acts of subagents. And an agent's own representation alone is not enough to establish his or her agency (*Home Owners Loan Corp. v. Thornbrush, Caprock Industries, Inc. v. Wood*). But if an agent is a general agent with the authority to hire subagents, the principal gives implied consent to the general agent's hiring choices. The authority for creating a subagent may come from an agent's written agreement with the principal or from the scope of authority that a principal gives an agent. A decision by a general agent to hire another agent for the principal could make the principal legally responsible for the actions of the subagent; if the hiring of the subagent fell within the realm of the general agent's duties as an arm of the principal. But when an agent is an independent contractor and hires his or her own employees to accomplish the work, the independent contractor, rather than the principal, is responsible for the employee's actions. For example, if a technical communicator who was an independent contractor hired a bike messenger to deliver materials to the principal and the bike messenger was injured while on the job, the technical communicator, rather than her principal, would have to pay the portion of the worker's compensation claims that could follow.

Principals' and Agents' Duties and Obligations

If you are a principal, you will be most interested in what an agent can do for you and how he or she can act on your behalf. The principal-agent relationship creates a special duty, called a "fiduciary duty," in the agent to the principal (Reuschlein 10–11). This means that the agent owes a special duty to the principal and must use special care to avoid creating any uncalled-for legal harm to the principal. A fiduciary duty is a higher-than-average duty of care created to benefit the principal. The agent owes a principal three main duties: obedience, care, and loyalty. For instance, a graphic designer who also helps make financial choices for a business and has access to the company checkbook has a special duty of care not to spend irresponsibly or to leave the checkbook in a place where someone might use it to write unauthorized checks on the company account. If the graphic designer wants an especially expensive piece of equipment for work but knows that buying it would put the company in financial jeopardy, even though the designer may have authorization to use the checkbook at his or her discretion, the designer has a fiduciary duty to refrain.

Agent's Obedience

An agent must follow the principal's directions and perform his or her duties in a way that best furthers the principal's goals, as long as those duties are consistent with the agreement that created the agency relationship. For example, if a technical communicator were hired as a general agent to produce a report on a new software application's use feasibility, the principal could direct, among other things, which employees to use for the project, which equipment to use, the date the project must be completed, and whether the technical communicator must work at the work site. But unless the agreement included a special arrangement to allow it, the principal could not also demand that the technical communicator complete purchase orders for new equipment for the work site, interview employees for positions in unrelated projects, or wash the company cars.

This arrangement protects the agent, but it also protects the principal. An agent must perform only those duties that the principal authorizes. Without authorization, an agent could not choose to send employees on a business trip in first-class seating if he or she were authorized only to book tickets in coach. In contrast, an agent is not bound to obey orders that are illegal or unethical, as in the case of a real estate agent who was pressured by his employer to engage in racial discrimination in showing properties (*Ford v. Wisconsin Real Estate Bd.*).

Agent's Care

Agents must use reasonable care and diligence to complete their duties to the principal. As a result of agents' special relationship with their principals, they have access to information that may be secret to the principal's business dealings, and which could be personal or volatile in one way or another. An agent must be careful to be circumspect about the principal's dealings in order to secure the principal's best interests. The agent must never use information about the principal for personal gain, to embarrass the principal, or to otherwise violate the fiduciary trust created by the agency relationship.

Agent's Loyalty

The principal has a valid legal expectation that the agent will be loyal by acting solely in the principal's interests. To that end, an agent must report the full amount of a principal's profits or losses, acting to minimize losses and maximize benefits to the principal's business. The agent must disclose information that might affect the principal adversely and must never undermine the principal's interests in favor of his or her own. For instance, if an agent who, by virtue of working in the principal's place of business, learns that a competitor is selling all of his or her business equipment in response to the threat of bankruptcy, the agent cannot make a personal bid on the equipment for personal gain, but must tell the principal about the sale instead.

Principal's Duties to the Agent

If you are a principal instead of an agent, you should be aware that you also owe a duty to your agent. The agency relationship is reciprocal. Where an agent has a fiduciary duty to a principal, the principal has a duty to an agent as well, although the higher fiduciary concern is not required. The principal's obligation to the agent is implied by law, in addition to what may be specified by contract. The principal has four general duties to the agent. The principal must (1) compensate the agent for the reasonable value of the

services performed, (2) provide the agent with the means to perform his or her duties (including necessary transportation, materials, work space, equipment, etc.), (3) protect the agent from liability
for the detrimental result of required actions in pursuit of the principal's goals and from expenses incurred while performing his or
her duties, and (4) the principal must not embarrass the agent, hurt
his or her reputation, or interfere in the performance of his/her duties to the principal. For instance, the personal information a technical communicator gives to a principal's human resources office
cannot be used against him or her for the principal's gain. For example, if the principal's writing team supervisor happens to review
the agent's employment file and learns that the agent was out of
work for three years, the supervisor cannot use the information to
make fun of the agent at the company party by calling him or her a
"Gen-X Slacker."

What Does an Agent Have Authority to Do?

What does an agent have the authority to do? An agent can make
business transactions that legally obligate the principal. Agency authority places the agent in the same legal relationship that the principal would have been had he or she made the transaction. The agent
obligates the principal to all the rights and legal responsibilities created by the arrangement. Explicit (actual) or implied authority to act
for the principal arises from either specifically stated (express) or implied delegation of duties. A principal may make specifically stated
(express) statements, often in written form, indicating exactly what
he or she expects from an agent. Additional implied duties arise from
what can be reasonably inferred from those express statements. For
example, if a multimedia designer, as an agent, has authority to use
the company computer by a specifically noted statement to that effect, an implied authority arises that allows the agent to use the software that operates the computer as well.

Agency authority is not absolute, as previously discussed, and
there are also distinctions in types of authority. Agents can act on
specifically stated (actual) or implied authority that arises from
specifically noted or implied duties. Actual authority allows an agent
to do what the principal specifically directs the agent to do. The principal may direct an agent through oral or written instructions, which
may include reasonably implied tasks that allow an agent to complete the job that he or she is directed to do. But an agent may also respond to an implied authority that arises from reasonable inference.
For instance, an agent may be told explicitly to create a brochure to be

distributed to potential buyers of the product featured in the brochure. In addition, among the implied duties, the agent can reasonably assume that if working with the publishing department of the company to print the brochure also requires distribution to complete the job, he or she must accomplish this task as well.

An agent's authority ends under five circumstances: (1) the agent has reason to believe that the principal no longer desires that he or she act on the principal's behalf, (2) the stated duration of the agent-principal relationship ends, (3) the agent either quits or is fired, (4) the principal's circumstances change to the extent that they indicate there is no longer a basis for the agency relationship (the house the agent was bound to buy for the principal is destroyed by floods), or (5) the law operates to invalidate the relationship (such as death of the principal or invalidation by an illegal act).

An agent may also have inherent (innate) authority to act on behalf of the principal, particularly if he or she is a specific kind of agent who has professional status to carry out specific business in which the principal has no special training, ability, or knowledge. A lawyer, for example, is hired for the very reason that he or she has training to make legal judgments and to draw up legal documents that will benefit the client. A technical communicator also has special training in document design and rhetorical communication that provides a basis for making professional judgments about how to develop effective documentation and communication. The authority to make decisions within the realm of knowledge covered by this special training is assumed in the agency relationships between professionals and their principals.

An agent's apparent authority arises when the principal acts or speaks in a way that makes a third party believe that the agent has authority to act for the principal. Under these circumstances, even when the agent knows that he or she has no authority (thus, paradoxically, no agency exists), the principal's words or actions can effectively bind the principal (again, paradoxically, creating an agency relationship). For example, a technical communicator's documentation firm hires a public relations agent and authorizes him to write advertising copy for the agency, but the principal indicates to a third party that the agent has a great ability with sales and "can pitch a documentation project so well that he can sell a client on any product," indicating that the agent has authority to sell a technical documentation project as well as conduct public relations. Although the agent has been given no actual authority to sell a documentation project to the client, his authority is apparent; thus, the agent could bind the principal to an agreement with the client.

If the case were changed slightly so that the agent had no express, implied, or apparent authority to sell an advertising campaign to the client, but the principal found it advantageous that the agent, even without authority, had done so, the principal may provide ratified authority that gives the agent retroactive authority to bind the principal to an agreement with the client. In essence, the principal agrees after the fact that the agent had authority to bind the principal to an agreement with the third party. But "ratification requires that the principal have full knowledge of all material facts concerning the transaction, and the principal must indicate through words or conduct that he or she intends to be obligated in the transaction" (Moye 9).

Differences in Legal Characterization: Independent Contractors and Employees

As you have read in the preceding information, agents and their principals have very different legal responsibilities to each other and to the people they do business with. Those responsibilities differ depending on agents' status as employees or independent contractors. Agents who are employees can obligate their principals (in this case, employers) to the legal responsibilities the employees incur. In contrast, independent contractors—rather than their principals—are legally responsible for their own legal engagements with others. In addition, by virtue of the work-for-hire doctrine—discussed in detail later in the book (see p. 97)—employers hold copyright interests in intellectual products created by employees who develop those products in the scope of their employment duties. They do not hold copyrights in work created by independent contractors unless they contract with each other to transfer the copyright. Individuals often find it difficult to determine the status of their relationships to their principals. When conflicts arise over their legal status and the duties that go with it, they resort to courts for clarification. Under earlier agency law (1909 Statute), the courts determined employee status based on the principal's right to control the activities of the agent, but today they consider the following 13 agency law factors in determining whether a creator is an employee or independent contractor.

- Does the hiring party have a right to control the manner and means by which the product or service is accomplished? The greater the extent of control a hiring party has, the greater the evidence will show an employer-employee relationship. For instance, if a principal hires a technical communicator to write a manual and tells her the subject to write about, which equipment to use to write the manual, and which graphics she must include, the court will find evidence of a right to control the manner and means of product development. This element, among others, can lead to a determination of employee status.

- What level of skill is required to complete the work for the principal? If it is necessary that an agent have a unique or a high level of skill, especially one about which the principal knows little or nothing, the court is likely to find evidence that the agent is an independent contractor. In part, it can be reasoned that a principal would have little control over what he or she knows little about, and would be less likely to determine the manner and means by which the product or service is completed. But it is important to remember that the right to control is no longer the only factor in making employee status determination. Another aspect of this evidentiary question goes to the likelihood that an individual with a unique skill or special ability or knowledge would be working independently rather than for an employer. In these cases, independent contractors often complete one-time products or services and do not have permanent arrangements with their principals; thus, they are less likely to be employees.

- Did the principal or the agent provide the instruments and tools used? In the case that an agent uses his or her own equipment and supplies to produce a creative product or service for a principal, the courts factor in favor of a showing of independent contractor status. If, for example, a multimedia designer created a digital film clip that the principal used to advertise a product, and the designer used his or her own computers, employees, and software to design the product, this factor would decrease the likelihood that the courts would find a showing that the agent is an employee. Again, this is only one of 13 factors, and if the courts found that every other factor indicated employee status, this one factor would be overridden.

This may be a good place to illustrate the importance of these determinations with another example. Consider a scenario in which a principal has hired an agent who is a multimedia designer to create a set of digital instructions for building a motorized go-cart. The agent uses the principal's equipment and creates the instructions under the principal's guidance and to his or her specifications. The agent has a moderate level of skill in developing multimedia products, but welcomes the principal's advice. After the designer produces the instructions, the principal distributes them, only to learn that the warnings and cautions are inadequate and that purchasers relying on the instructions are being hurt in the process of building their go-carts. The principal is facing numerous lawsuits based on claims of negligence in selling products with faulty instructions. The principal claims that the multimedia designer was at fault and should carry the liability for the negligence claims, but the designer claims to be an employee rather than an independent contractor, and therefore that the principal bears the legal burden of the designer's actions in service to the principal. Based on the facts of this scenario, the courts are likely to find that the elements described, if consistent with a conglomeration of like findings in the factors to be explained next, show that the multimedia designer is an employee and that the principal, rather than the employee, is liable for the designer's actions in creating faulty instructions.

Other factors that the courts consider in determining employee or independent contractor status follow:

- Did the agent work at the principal's or agent's place of business while creating the product? If an agent works at the principal's place of business, there is a greater likelihood that he or she will be determined to be an employee rather than an independent contractor. Traditionally, independent contractors work at home or in their own workplaces. However as telecommuting becomes more feasible, acceptable, and popular in work relationships, this element will become difficult for courts to use as a factor for determining employee status. For this reason alone, courts will not be likely to place great emphasis on this one element for making their determinations.

- What is the duration of the relationship between the principal and agent? When agents have worked for principals over long periods of time, as for several months and particularly for years, courts note this as evidence leading to determinations of employee status. Independent contractors often work for principals for short periods that are only long enough to produce work on a project-by-project basis. Employees, in contrast, work over long periods, often working from one related project to another without ever completing a term of employment. When courts see an established pattern of a long-term work relationship rather than a series of short-term project completion arrangements, they are likely to apply this factor toward finding that the agent is an employee.

■ Does the principal have the right to assign additional projects to the agent? In another aspect of "the control of the worker" test, if a principal has the right to assign additional projects to the agent without formulating new contracts with each assignment, courts find evidence of employee-employer relationships. The implication is that the agent is working with the principal for the long term and is hired to continue working from project to project rather than contracting for a single term of work, like an independent contractor. If you work as a technical communicator in a business that handles multiple technical communication projects or that needs multiple forms of document development to support its work, such as that in an engineering firm, you may be assigned numerous projects over time. If this is the case, this element of your relationship with the employer would lead to a work-for-hire employee status determination.

■ How much discretion does the principal have over when and how long the agent works? When a principal can set an agent's work hours and determine when and how long he or she works, this manifests control but also indicates that the agent is not independent from the employer and works, instead, under the principal's direction and management. Courts will be more likely to find that a worker is an employee when his or her work hours are controlled by the principal.

■ What method of payment does the principal use to pay the agent? When principals pay agents in one lump sum for their work or in several partial payments for one work product, the agents are likely to be considered independent contractors. But agents who are paid weekly or monthly for continuing work are generally more likely to carry employee status. If you are a web designer who is paid for one site design after another on a piecemeal basis, you will be more likely to be determined an independent contractor. If you work as a web master and constantly update and redesign the same web site, you are more likely to be determined an employee.

■ Does the agent or the principal hire and pay assistants? Agents who hire their own assistants demonstrate that they control their own affairs and show independence from their principals. Based on this element, they are less likely to be considered employees. Remember that one element is not an absolute determining factor of employee or independent contractor status, however. There are times when employees do have the responsibility to hire assistants but are still considered employees. For example, team managers in advertising firms, technical communication reporting teams, and multimedia development partnerships, among others, may hire assistants to aid production but may also use equipment and materials supplied by the principal, work on-site at the principal's workplace, be paid on a continuous basis, and be supervised by the principal. The synthesis of all these elements can lead to a showing that these managers are, after all, employees.

- Is the work a part of the principal's regular business? When a principal hires an agent to work of which the nature is part of his or her regular business, the court will find a greater likelihood that the agent is an employee than if the agent were hired to perform an unusual service or create an unusual product. An engineering firm manager might hire a technical communicator to help write a report on the efficacy of a new jet propulsion system created by the firm. This work would be consistent with the principal's usual course of business. But if that same manager hired a multimedia designer to create digital family photographs, the court would likely find that this was not a part of the regular course of business and that the agent would be more likely to be an independent contractor.

- Is the principal in business at all? If the principal is not in business at all, courts may find it difficult to establish an employee-employer relationship. A principal who has no ongoing business is much more likely to have established an independent contractor relationship with an agent, who, by virtue of having no means to work for the principal for more than on a one-time or project-to-project basis, is likely to be an independent contractor.

- Does the principal pay employee benefits to the agent? If the principal pays retirement, medical, or other benefits to the agent, the courts will factor this behavior into finding an employee relationship. Independent contractors most often receive no benefits from their principals.

- How is the agent treated for tax purposes? If the agent is taxed as an employee, courts will use this element to contribute to a finding of employee status.

It is not necessary to meet each of these 13 criteria to be considered an employee. Instead, the courts base their decisions on a conglomeration of the elements to decide whether a person is an employee. But in making best assessments of your current status, these elements should guide you. In nearly all cases, it is best to create contracts that clarify your rights and duties so that you know what to expect.

Business Organizations

Agency relationships are the most basic structures for business relations. But technical and other creative communicators may wish to develop more complex forms of business enterprises and will find it helpful to understand the differences among sole proprietorships, general partnerships, limited partnerships, limited liability corporations, and corporations. These business organizations provide structures that determine the extent and kinds of legal liabilities shouldered by the individuals who make them up. Business structures also determine the roles of their various participants and set a basis for establishing the tax consequences of the enterprises they undertake. (Note that further detail on the business organizations that follow can be found in John Moye's *The Law of Business Organizations* and Larry Ribstein's *Unincorporated Business Entities*).

Sole Proprietorship

A sole proprietorship, also called an individual proprietorship, is the simplest and most common business organization. In this structure, one individual owns all the properties of the business and pursues all the work involved in operating, controlling, and developing creative products or services for the business. A sole proprietorship operates as an extension of an individual. For instance, a technical communicator might establish an editing business as a sole proprietor. She would own and control the equipment and materials used, would own the business property, and would also complete the editing work for her business's clients. Where she as an individual might work as an editor, the editing business—a sole proprietorship—is an extension of himself or herself.

The advantages of a sole proprietorship include the ease by which it is formed and its inherent operating flexibility. There are no special legal requirements for a sole proprietorship and no need for filing legal papers to support it. Sole proprietors can handle their business as they choose, hire the employees they wish to hire, and, among other things, set the working hours and conditions any way they please.

The greatest disadvantage of a sole proprietorship is that the business is completely identified with the owner, who is individually

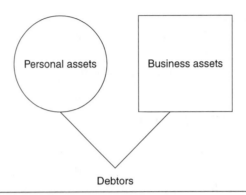

Figure 3.1   Debt carrier's reach to sole proprietor assets

liable for any misactions that result from business operations. The business is founded wholly on financial responsibility of the owner, through his or her personal contributions. There are no stocks or shares to support the business, and all loans and debts are the personal responsibility of the owner. Sole proprietors often have limited ability to borrow funds when necessary to support their business.

The sole proprietor's personal responsibility for their business creates an unlimited legal responsibility for any debts, negligent acts, taxes, or harm resulting from business operations. This means that the sole proprietor's personal assets are legally reachable to pay legal claims. Sometimes sole proprietors are able to buy insurance that will help protect their assets, but essentially, sole proprietors are responsible for every event that occurs in business in the same way that they are responsible for results of their personal dealings.

Although there are very few exceptions, a sole proprietorship terminates only upon the death of the owner.

General Partnership

You may find that you want to work with one or more partners rather than working alone. This arrangement is often advantageous when partners have varied skills and can create a more robust business together than they could do individually. When more than one person create a business together, they make up a general partnership. A general partnership is distinguished from a sole proprietorship in that it employs more than one owner. It is an association of two or more people whose purpose is to establish and operate a business in order to create profit. A general partnership is usually formed on the basis of an agreement—a contract between or among the partners that specifies their roles and responsibilities in the partnership business. Just like employee-employer and principal-agent relationships are agencies, a partnership forms agency relationships between and among the partners and any employees or independent contractors who work for the business. The Uniform Partnership Act defines a partnership as "an association of two or more persons to carry on as co-owners of a business for profit" (UPA §6 (1)).

A partnership agreement is required to form the partnership, but it is not necessary that it be in writing. Nevertheless, the partnership agreement is the most important element for forming a partnership because it governs the rights and responsibilities of the partners, outlines the purpose and structure of the business, and clarifies other terms of the agreement to operate a business together. The agreement usually notes the percentage of profit and loss sharing as well, but if it is not clearly laid out, the partners share all profit and loss equally. Ideally, this agreement should be clear to all parties and should be in writing to maintain clarity in regard to duties and responsibilities as memories fade.

A general partnership operates much like a sole partnership in terms of allocation of personal liability of debt, tax responsibilities, and legal claims. The main difference between the legal responsibilities of a sole and general partnership is that in a general partnership all the partners are personally legally responsible for the debts of the partnership. In addition, all the profits of the business are allocated to each member of the partnership and each is legally responsible for his or her share of tax.

Forming a general partnership does carry some advantages. This form of business organization is simple and easy to form and can be created quickly. Where the partners maintain a trustworthy relationship, business costs are spread among partners, making them easier to bear, and the potential for profit is often increased when costs and responsibilities are divided between or among partners. Relative to dealing with corporate arrangements, managing a general partnership is uncomplicated, and it is fairly easy to oversee business dealings as they develop, giving each partner a clear insight into the business's operations.

The greatest disadvantage of a partnership arrangement is that each partner is responsible for losses or legal claims on the business, even when another partner fails to dispense with his or her obligations. The following scenario is, unfortunately, more common than it should be among partnership organizations. Roy and Abel form a partnership to provide instructions for finding and restoring vintage cars to resell for profit. Their partnership agreement provides that all profits and losses are to be shared. The profits are shared equally, and at tax time, Roy declares his profit from the business and duly pays the required income tax. Abel, on the other hand, does not pay the tax on his half of the profits and neither notifies Roy that he has not paid, nor notifies him that the IRS has threatened to levy Abel's bank account to withdraw the required tax Then the IRS, finding that Abel has no money in his bank account and without notice to Roy, levies Roy's bank account instead, withdrawing the required tax. Partners in a general partnership are responsible for the legal responsibilities of the other partners when they result from operations of the business, and have to recoup losses from the failing partner rather than from the business's creditors.

Partners cannot protect themselves from the IRS, but they can protect themselves from legal responsibility to other people or business organizations that do business with the partnership (third parties) if they negotiate agreements with those parties prior to doing business with them. Of course, third parties are often wary of making these kinds of agreements, and negotiate the terms carefully. But any time there is doubt whether business dealings will operate smoothly and fairly, creating agreements to clarify terms can help alleviate strain and potential legal problems.

As is the case in sole proprietorships, if a party to a general partnership dies, the partnership is dissolved. A partner's retirement or some other form of disassociation from the partnership will also dissolve the partnership. But these means of dissolution can be overcome if the partners make other arrangements in the partnership agreement. The remaining partners can continue doing business as a partnership as long as the agreement makes it possible. Making arrangements for death, retirement, or disassociation can be important in a business that has built up value in good will, the intangible but economically valuable asset that represents a company's good relationship and good standing with the public. Intangible business assets may be more valuable than tangible ones, and it may be more economically advantageous to partners to continue working in a business enterprise than to sell it.

The partnership organization itself pays no federal tax, but each partner must pay income tax on his or her share of profits. All the losses of a partnership are attributed to each of the partners in shares. Where the agreement makes no statement about how losses are to be divided, they are proportioned to mirror the percentage of gain each

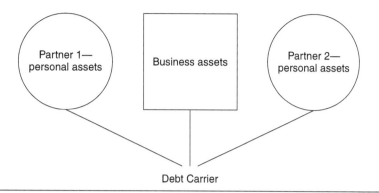

Figure 3.2 Debt carrier's reach to general partnership assets

partner would make. But partnership agreements can lay out any terms agreed on by all the partners.

Limited Partnership

You may be in a situation where you either work with or are a limited partner in a business characterized as a limited partnership. A limited partnership is a business arrangement, established to make profit, that incorporates the contributions of both general partners and limited partners as co-owners of the business organization. Most states have adopted the Revised Uniform Limited Partnership Act. In all states, however, a limited partnership provides the limited partners with an arrangement similar to what they would experience in a corporation. And in all states, limited partnerships can be formed only with a formally prescribed statutory arrangement and cannot be formed through a private agreement among the partners.

Limited partners participate in the limited partnership much as they would in a corporation. Limited partners' legal responsibilities extend only as far as the contributions they make to the partnership, so their personal assets are protected from legal responsibilities that the partnership incurs. For example, a limited partner might contribute $50,000 to establish a media design business entity. Two general partners might contribute equipment, time, and effort and $5,000 each to the business. If the partnership fails to achieve its financial goals and has a loss of $70,000, the limited partner would forfeit only his or her $50,000 contribution, and would not be legally responsible for any more of the loss. The general partners, however, would lose their $5,000 each and would also be responsible for the remaining $10,000 loss. If pushed to the limits of the partnership agreement,

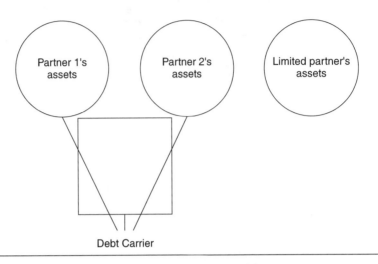

Figure 3.3 Debt carrier's reach to limited partnership assets

however, a limited partner, unlike a shareholder in a corporation, could eventually be held responsible for debts to creditors if the partners failed to provide compensation to creditors.

As the example above illustrates, general partners bear the extent of the risk of legal responsibility that results from the partnership. As is the case for sole proprietors and partners in a general partnership, general partners' personal assets are vulnerable to creditors, the IRS, or those who have won legal claims against the business. A limited partnership requires that at least one partner be a general partner who will bear the risk of doing business. Appropriate to the extent of legal responsibility borne by the general partner, he or she also maintains management control and responsibility for the business's affairs.

Limited partners are prohibited from participating in the partnership in any form, for any participation would undermine their limited status. But even general partners have some limitation on the extent of their actions in the name of the business. They may not act in a way that is inconsistent with the partnership agreement, and they may not act, in the name of the business, in ways that do not support the purpose of the business.

In addition, general partners may not interfere with partnership business, use the partnership property for personal or other than partnership purposes, or admit other partners into the operation of the business. Fundamentally, the general partner has a fiduciary duty (remember that this is a special duty of care) to the other partners and must work to the best of his or her ability to support partnership interests.

Getting Out of or Ending a Partnership

General partners can withdraw from a limited partnership either intentionally or by accident. The terms of the partnership agreement affect what happens when a general partner withdraws from the partnership and, in cases where there is a penalty for withdrawal, also dictate the fine or other form of penalty a withdrawing partner might have to pay. A general partner can withdraw by giving notice to the other general partner(s) and to the limited partner(s), but again, he or she may have to pay a form of fine (damages) for the withdrawal.

A partner may also assign (legally transfer) his or her interest in the partnership to someone who is not a partner. This will entitle the person or business organization that receives the partner's interest to the same portion of profits or loss in the business that the general partner would have had. This transfer of interest is another form of the original partner's withdrawal from the partnership. The partnership agreement may also provide a list of circumstances that would justify a partner's removal, and, when the circumstances exist, the partner can be withdrawn in response. For example, an agreement may provide that the general partner in a technical communication business must make timely payments to the business property mortgage holder. Over time, the general partner's failure to make payments on the property leads to foreclosure and the business loses the property. The partner can then be withdrawn from the partnership and also may be legally responsible for the losses resulting from the foreclosure.

In addition, unless the partnership agreement excuses it, partners may be withdrawn if they admit to personal insolvency, making them incapable of bearing their portion of financial responsibility for the partnership. And as it is with the sole proprietorship and general partnership, a partner's death or legal incompetence can lead to withdrawal. A general partner's withdrawal dissolves the partnership the same way it does in a sole proprietorship or a general partnership, with three exceptions: (1) all the partners consent that the withdrawing partner may continue, despite the occurrence of dissolution events, (2) another general partner may act in the withdrawing partner's place, or (3) all of the parties to the agreement may agree in writing to appoint one or more additional general partners to manage the business.

Unlike a general partner, a limited partner may withdraw from the partnership without dissolving it. And the limited partner can also withdraw and regain his or her contribution as long as the partnership agreement has no restriction on this ability. In addition, limited partners can be added to the partnership without the consent of other limited partners. This potential is dependent, however, on the conditions that all partnership creditors have been paid and that the partnership's assets are sound.

A limited partner's actions can also dissolve a partnership. The same conditions that apply to ending a general partnership apply to limited partnerships, but a limited partner's misconduct also can dissolve a partnership. In addition, a limited partner can ask for a dissolution when it is no longer practical to carry on business.

The tax liabilities of a limited partnership operate the same way they do in general partnerships. Each individual is responsible for taxes on his or her profits from the business. Often, limited partners enter partnership arrangements in order to offset profits from other passive income by benefiting from tax decreases resulting from losses in a less profitable business. It is not uncommon for limited partners to intentionally participate in unprofitable business ventures as a way to save more money by decreasing tax liability than they lose by contributing to unprofitable business ventures.

Limited Liability Corporations

Limited liability corporations, commonly called LLCs, are a relatively new form of business entity and have become very popular, particularly with the rise of entrepreneurial dot.com businesses. All states now have a limited liability company act that provides regulations for forming an LLC. An LLC may be thought of as a hybrid of a partnership and corporation. LLCs are managed much like partnership businesses, but members are also able to limit their legal responsibility the same way they can within corporate structures. LLC tax structures operate in the same way as S corporations do in allocating taxable profits only to shareholders and not to the LLC. In other words, they are often taxed like partnerships—with the exception of those in Florida, which are taxed like corporations for state tax purposes. All LLCs are different from corporations, however, in that there is no requirement for board meetings, boards of directors, officers, or stock certificates. They are also different in that there is no requirement that shareholders meet S corporation qualifications.

Any number of members can be joined in an LLC. A group of members may also jointly become LLC shareholders of a single share or set of shares. In addition, all LLC members are granted limited liability protection from business debts the same way that limited partners are allowed an element of personal freedom from legal responsibility for limited partnership debts.

An LLC is a legal entity that is separate from the owners who maintain a membership interest, and it is able to exercise statutory powers similar to those of corporations. LLCs are legally "anthropomorphized" in the sense that the LLC can act, through legal construction, in the same way that individuals can act to carry out business or other LLC affairs. Through this legal anthropomorphization, LLCs can also own stock or a percentage of stock in a company or companies. LLCs can participate in nearly any business function. LLCs' functions are restricted to those granted through the statutes of the state in which the LLC is created, along with the articles of organization formulated by the LLC's members. For instance, a multimedia designer, artist, technical communicator, graphic designer, and non-participating investor decide to form an LLC in order to do business as a multimedia firm. They own the business but the LLC's assets, rather than their own, are subject to legal liability for the actions of the business. They feel safe in doing business as an LLC, knowing that starting up a dot.com business is risky. They form the LLC to protect themselves from leaving their personal finances and property open to legal liability.

LLC owners, called members, can contribute tangible or intangible assets the same way business partners can, and some may be passive investors who act very much like limited partners in limited partnerships. They may contribute their share of money, materials, or equipment, and then share in business profits without ever participating in business operations. Members do have statutory rights to vote, however, and managers of LLCs usually do the work of running company business. An LLC may have many managers or only one, depending on the articles of incorporation. Members have the right to approve amendments to the articles of incorporation, to admit or deny admission for new members, and to continue the business at the point of disassociation, if all members agree. Members can also assign or transfer their memberships in LLCs, but in some states, all members must agree to assignation or transfer. Normally, a purchaser is not allowed to participate in management or become a member but will only have rights to financial benefits.

LLCs usually have a statutory limitation on duration of operations for 30 years, although this may vary from state to state. At the end of this term, LLCs would dissolve. Withdrawal of any LLC member will also dissolve an LLC, as will an agreement made by the membership, provided that all debts are paid to creditors. Although there are statutory variations, other catalysts for dissolution include retirement, resignation, bankruptcy, and death or incapacity.

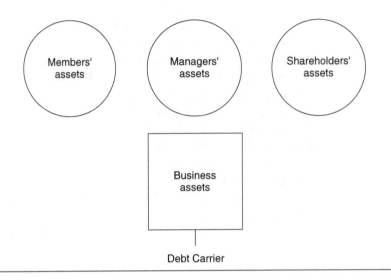

Figure 3.4   Debt carrier's reach to LLC and corporation assets

Corporations

Corporations are extremely complex legal entities that can only be created through statutory development. A corporation is treated as an individual legal entity, as a "person" under the law. Its existence is controlled by state statutory regulations, its articles of incorporation, and bylaws that lay out its operating framework. Its owners act as shareholders, directors, and officers who operate the business, but the corporation itself owns the business. It exists as an entity separately from its shareholders and managers. Shareholders can buy assets in the company without owner approval and can transfer shares to others with no effect on the corporation's legal status.

A corporation is liable for its own financial obligations and liabilities separate from its owners, whose individual assets are usually not reachable. Shareholders elect managers to control the operations of the corporation. Shareholders, owners, and managers all have limited legal responsibility for the corporation. Generally speaking, corporations are taxed like a natural person, but S corporations, or shareholder corporations, are taxed like partnerships. I have provided merely a snapshot of corporate description here. The process of forming and maintaining a corporation is extremely complex. If you consider this option, you should certainly use the services of a lawyer and an accountant who are well versed in corporate law.

You can see that there is a broad variety of working relationships and legal arrangements among and between people doing business. Ideally, before entering into relationships with others, you will use this information to help you understand the potential consequences of your business relationships at work. There is never a guarantee that you can avoid legal conflict altogether, but understanding the basic structures that control your work will help you make sound judgments. As with all other areas of law, if you foresee the possibility that you might encounter a particularly intricate or difficult situation in the future, or if you already have, the best step you can take is to consult a lawyer whose focus is the specific area of law where you are encountering difficulty. In that case, the material here will help you make informed choices about how to proceed on your lawyer's advice.

Discussion Questions

1. What are some examples of how each of the 13 agency partnership elements determining employee status for work-for-hire purposes plays out in your own workplace experience?
2. What are the relative advantages and disadvantages of employee status?
3. How could you benefit from each of the business entity arrangements described here?
4. What would be your legal liabilities in each of these settings and the relative benefit and detriment of taking them on?
5. What are the relative advantages and disadvantages of each of the categories of business entities?
6. At what point would you consider hiring a lawyer to help you with your legal relationships? Why?
7. How far does fiduciary duty to an employer extend? To a friend? To a family member? Please explain your answers.
8. Under what circumstances would you consider starting a sole proprietorship?
9. Would you consider entering a general partnership? Limited partnership? LLC? Corporation? Why or why not?
10. Have you ever been in an agent/principal relationship? What were the circumstances? Were you aware of the legal ramifications of the relationship?
11. Now that you have a greater understanding of the laws that govern business relationships, you can use the brainstorming document that you created after reading Chapter 1 as a basis for considering options for business arrangements. To do this, write an analytical report based on your goals and requirements as developed in your

brainstorming document. The report should provide an accounting of your business goals and an analysis of the relative benefits and detriments of each form of business entity that you might use to meet those goals. In another section, you should provide a listing of reasons explaining why some of the potential business arrangements should be eliminated from your choices and why others might be reasonable choices to make. You should conclude the report by noting a final choice and giving reasons for that choice.

Legal Agreements: Contracts

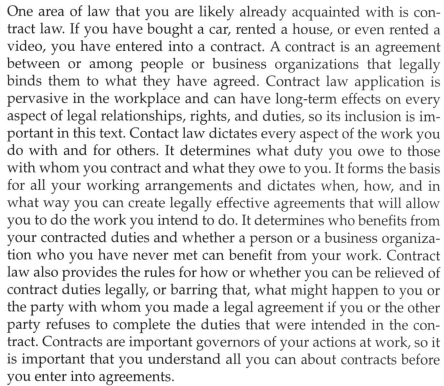

One area of law that you are likely already acquainted with is contract law. If you have bought a car, rented a house, or even rented a video, you have entered into a contract. A contract is an agreement between or among people or business organizations that legally binds them to what they have agreed. Contract law application is pervasive in the workplace and can have long-term effects on every aspect of legal relationships, rights, and duties, so its inclusion is important in this text. Contact law dictates every aspect of the work you do with and for others. It determines what duty you owe to those with whom you contract and what they owe to you. It forms the basis for all your working arrangements and dictates when, how, and in what way you can create legally effective agreements that will allow you to do the work you intend to do. It determines who benefits from your contracted duties and whether a person or a business organization who you have never met can benefit from your work. Contract law also provides the rules for how or whether you can be relieved of contract duties legally, or barring that, what might happen to you or the party with whom you made a legal agreement if you or the other party refuses to complete the duties that were intended in the contract. Contracts are important governors of your actions at work, so it is important that you understand all you can about contracts before you enter into agreements.

Chapter 4 explains issues in contract law, defines the basics of the three elements that make up a contract—offer, acceptance, and consideration—and notes who can and cannot enter into contracts.

This chapter also explains many of the potential effects of contractual obligations, defines breach of contract (breaking the agreement), and explains the potential effects. You will see that issues in contracts, like other areas of law, are based on your relationships with others.

As you have seen in the preceding chapters, the law provides a framework of answers to the problems that can arise during everyday activities at work. But you have also seen that the answers are always relative to the context in which legal conflict arises and that they rarely provide enough certainty to enable you to rely on legal patterns for planning your future in business. In fact, while there is no area of the law that provides absolute answers to potential conflicts, one way to help clarify your legal duties and rights in your relationship to your employers or employees is to create contracts. As you will learn in this chapter, even a contract will not give you absolute certainty about the legal outcomes of your business dealings, but it can help you gain as much certainty as possible before you begin your work. It also can make clear from the beginning of your dealings with others what they expect to provide you and what they expect you to provide them. [Note that West's Handbook Series provides very detailed material on contract law in John Calamari and Joseph Perillo's *The Law of Contracts*.]

Creating a Contract Through an Offer, Acceptance, and Binding Exchange

Although written contracts are always preferable, since they help support one's memory of a contract's terms, with few exceptions, a contract can be valid whether it is oral or written. Basically, a contract is an agreement between two or more parties that is legally enforceable. It is a bargain between two or more "persons," who can be real people or, as you might remember from the chapter on agency, business organizations acting as "persons" under the law. The contractors may bargain for a promise, for an act, or for avoidance of an act. Contracts for sale of goods are covered by the Uniform Commercial Code (UCC), adopted by nearly all states, but all valid contracts—commercial or otherwise—require three elements: offer, acceptance, and consideration.

These three elements are the foundation of all contracts. The "offer" is the suggestion provided by the offeree (the person who makes the offer) to bargain with the offeror (the person who can enter the bargain and accept the offer). "Acceptance" is the offeree's agreement to enter the bargain by doing or forgoing some kind of action, or by paying an agreed upon sum of money or providing some other form of benefit, as the offeror has asked. It can take the form of a promise to act, to pay, or not to act, or the action itself, payment itself, or forbearance of action. "Consideration" completes the contract. For instance, an owner of a technical communication firm might ask a multimedia designer to create a digital video for teaching employees how to operate the new paper-binding machines. This action would be considered an offer, the first element of a contract. The owner then tells the designer that compensation would be in the form of both a set fee and a share of stock in the company. This second element of the contract, the fee and stock, are the requisite consideration. At this point, if the designer declines to do the work because he would rather not take the stock, the three elements of contract would not be met and there would be no contract. But if the designer agrees and decides to do the work for the stock and the fee, then his or her acceptance will have fulfilled the last requirement for creating a contract.

In addition to the parties providing the three requirements of offer, acceptance, and consideration to form a contract, they also must have the intent to create a contract. They must know that they are contracting, be capable—both legally and mentally—to contract, and they must have a common idea about what they are contracting for—called a "meeting of the minds." In determining whether there is a meeting of the minds to form a contract, the courts often question whether it is reasonable to believe that one party was making an offer to another and then whether it is reasonable to believe that the other party was accepting. For instance, the courts will decide whether the parties were serious about entering a contract or only joking. The parties must intend that the contract is legally enforceable and their intent makes the contract legally enforceable. If both parties desire a contract to be legally enforceable, it will exist even if it comes from a mistake. For instance, if both parties desire to be bound by an oral contract, a court can enforce it, even if the law states that the contract must be in writing to be enforceable. Where there is a question as to whether a contract exists, courts begin deciding the question by basing their inquiry on the presumption that there is a contract. For example, if a digital film developer's acceptance of the offer to design a digital video for a fee and company stock is based on a misunderstanding of what was meant by "digital video," then there would be no meeting of the minds and there would be no contract.

Proposing an Agreement: Offer

An offer is a proposal by one party to another to enter into a bargain where each party provides something to the other. An offer indicates a willingness to enter into a bargain and creates the power for the other party to accept. A legally valid offer has to be based on the real intent of the offeror. An "offer" in the form of a joke is not considered legally valid, but the contracting party has to know that the joker is not serious (*Mears v. Nationwide Mut. Ins. Co., Lucy v. Zehmer*). Preliminary negotiations such as bid soliciting for construction jobs are not considered offers, but rather a means to decide which bidder a potential contractor will actually contract with in the future. Most newspaper advertisements are not considered offers because they are usually not clear on quantity, quality, and other specific descriptions or are too vague to be considered offers. But if an advertisement includes specific terms, such as "Patagonia cotton turtleneck sweaters, $26 each, first come, first served," or if there are specific words of commitment, such as "Will sell, without reserve," then a valid offer will have been made.

Accepting to Enter an Agreement: Acceptance

Acceptance is a showing of agreement to be bound by the terms of an offer. Only the person with whom the offeror intended to bargain can accept an offer. If Joe makes an offer to Kerry, only Kerry can accept, even if Jane likes the offer and wants to accept it. Joe would have to make an offer to Jane in order to create her "power of acceptance." The offeror also decides the form that contract acceptance must take. For instance, the offeror may state that the offeree must accept by E-mail. If the offeree attempts to accept with a phone call instead, the acceptance would not be valid. Where the offeror does not specify a method of acceptance, an offer can be accepted in "any reasonable" way. In fact, every aspect of contract law follows the objective theory of contracts or "the reasonable person test." In other words to determine contract validity, the courts look to what a reasonable person would expect the intent of the other parties to be.

An offer calls for the offeree to provide a promise or some kind of "performance," which is an action or forbearance of action for which the offeror asks. But where it is not clear whether the acceptance can be a promise or a performance (action or inaction that was asked for), either one is counted as acceptance. For instance, a manager of a dot.com communication development company offers to pay a graphic designer to create a set of images to be used as clickable icons for web links. The manager does not specify whether the designer should promise to create the icons or whether he should go ahead and create them. The designer can accept the offer either by promising to create the icons or by creating them. But if the designer, instead of creating the graphic icons, develops sound-based links, his action will not be considered acceptance because what he developed varies from the manager's specific instructions for acceptance. The designer's act operates as a counteroffer, which the manager can accept or reject. The manager who now has the power of acceptance (the power to accept), is not bound by either the original offer or by the designer's counter offer. If the designer created all the graphic icons that the manager asked for in response to the offer, the manager's action would be "unilateral acceptance" of the offer. Once the designer started work on the graphic icons, the manager could not "revoke"(take back the offer and dissolve) the contract.

Fundamentally, when a form of acceptance varies from what is asked for in the offer, it does not act as acceptance. If the acceptance is not the precise mirror image of the offer, it is considered a rejection. Nevertheless, UCC law that governs commercial sales—the sale of goods as a basis for a business for profit—rejects the mirror image rule in order to avoid a situation (called "battle of the forms") that allows a party to later wriggle out of a bargain that he or she intended to make. The UCC provides *generally*, that an "expression of acceptance"or "written confirmation" will act as acceptance even if the terms are "additional to" or "different from" the offer (UCC § 2-207 (1)). So where a seller offers a computer at an agreed-upon price and a specified delivery date, if the delivery date is slightly different and the difference in delivery does not have an extreme negative effect on the buyer, the acceptance will still be considered valid.

Generally, a person cannot accept an offer by remaining silent and not giving some more obvious indication of acceptance. But if it is clear to the offeree that the offeror will consider the silence an acceptance, silence can operate as such. If the offeree has benefited from action that constitutes an offer and makes no effort to reject the offer (or the merged benefit), this form of silence can be considered acceptance. For example, a technical communicator has been offered a chance to use a business's computing equipment in exchange for editing work on its annual report. The technical communicator is silent regarding acceptance but is able to contract with another company to write a brochure for it, based on the knowledge that she will have access to the equipment necessary to do the work. She does not reject the business's offer of equipment, and she has benefited from the offer even though she is silent. Her silence can be considered acceptance.

In your work in developing creative communication, you may have already encountered, or may encounter in the future, some situations where you want to accept an offer, but you would like to add terms to the acceptance. As long as neither you nor the offeree are selling and buying products as a business (merchants), your acceptance will be valid, but the offeree will have to explicitly agree to the terms. For example, a digital film developer in a communication firm accepts an offer to make a video clip that advertises a new car. The filmmaker adds to the acceptance a clause noting that all legal conflicts will be handled by an arbitrator rather than the courts. If the offeror explicitly agreed to the addition, the filmmaker's acceptance would be valid. The exceptions to the rule that an offeree may add terms to an acceptance are situations in which the changes alter the contract to the extent that they change the intent, or in which the offeror objects to the additional term.

In order for an offeree to accept an offer, he or she must know of the offer. It is impossible for someone to accept an offer that he or she is unaware of, because acceptance requires an intent to accept, and that is exactly the reason acceptance must be based on an offeree's knowledge of the offer. However, cases where there is disagreement in this area are more frequent than you might expect. Such cases most commonly occur when a reward is offered for an act and a person completes the action requested in the post of reward, but does so without knowledge of the reward. When the actor learns of the reward after completing the action, despite his or her belief that the reward is due, the law will not uphold a claim on the reward.

An offeree can accept an offer only as long as his or her "power of acceptance" (power to accept) is intact. In other words, if an offeror makes an offer (giving an offeree the power of acceptance) and sets a time limit within which the offeree must accept or reject the offer, the power to accept the offer lasts only within the time frame set by the offeror. Once that time is over, the offeree has no more power to accept. If the offeror sets no time limit, the power to accept nevertheless expires within a legally "reasonable" period of time. The power to accept can be ended in five ways: The offer can be rejected, the offeree can make a counteroffer, the time frame in which the power to accept is still alive expires, the offeree or offeror dies or becomes incapacitated, or the offeree validly takes back (revokes) the offer any time before acceptance. So, for example, if a school offered a technical communicator an acceptable fee to write a report on teacher effectiveness and gave him two weeks to make a decision about whether to accept the job, he would have to accept the offer within the two-week time frame for his power to accept to still be intact. If he tried to accept the offer after the two-week period had ended, he would have no power to accept and form the contract.

An offeror cannot "take back," or revoke, an offer until the offeree knows about it and if a letter indicating that the offeror takes back an offer (a letter of revocation) is lost in the mail or through a problem with E-mail, it never becomes effective. In addition, there are cases where offers are irrevocable, such as standard option contracts, where the offeror grants the offeree an "option" to enter the contract. If an offeree has already completed the action required by the contract or incurred some kind of debt, paid money, or incurred some other kind of detriment in reliance on the contract, the offeror cannot revoke the contract. These actions make an offer irrevocable, when the offeree actually performs the actions bargained for rather than just making preparations to complete his or her duties under the contract. You may remember that "performance" refers to acts that can constitute acceptance of an offer.

"Detrimental reliance" refers to situations in which an offeree, in response to an offer, undertook an action, forbearance of action, or paid money as a form of acceptance, leaving him or her in a diminished state. Detrimental reliance that is elicited by an offer obliges an offeror to complete the bargain with the offeree. For example, a bank asks a technical communicator to design an annual report. The technical communicator turns down another job where she has been asked to develop a web site. The technical communicator relies to her detriment by turning down another offer to design the website. By notifying the bank that she would design the report, the technical communicator responded in a way that acted as acceptance of the offer and in turn, lost her opportunity to design the web site, so she has a valid claim of "reliance" and the offer would be irrevocable. Had the technical communicator only indicated her desire to build the web site but was never offered the opportunity, she would not be in a position to claim detrimental reliance.

Offers by subcontractors to general contractors are temporarily irrevocable, in part because contractors can only bid on their general contracting jobs based on the figures for subcontractor bids. So where a contractor's bid depends on a subcontractor's agreement, she detrimentally relies on the subcontractor's promises to make her general bid possible.

When Acceptance of an Agreement Becomes Effective

You have seen that the law provides a set of rules for determining what can be considered acceptance and for how long an offeree has the power to accept an offer. Another area important to validating a contract is *when* the acceptance becomes effective. The "mailbox rule" is that acceptance is effective upon proper dispatch, so if a letter is

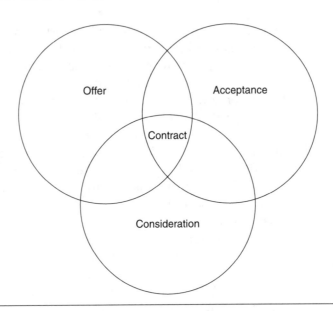

Figure 4.1   Elements of a contract

considered valid acceptance, it becomes effective as soon as the letter is dropped into the mailbox. Likewise, if E-mail is considered a valid means of accepting an offer and the offeree sends E-mail accepting it, the acceptance would become valid as soon as the E-mail was sent. Remember that an offeror has the power to decide the form an acceptance must take, so if an offer provides that the offer is accepted if and when the acceptance is personally received by the offeror, then the mailbox rule does not apply. If an acceptance is lost or delayed, applicability of the mailbox rule depends on whether the letter was properly addressed, in which case the acceptance period would last until the end of the dispatch, even if the acceptance was lost and never received. If an acceptance was not properly addressed or dispatched, it would be effective only if it were received within the time frame that a properly dispatched acceptance would normally take to be received. So if a computer company manager offers an equipment purchaser a digital communication system for her communication firm, indicating that E-mail acceptance would operate to create the contract, and if her server shuts down when the E-mail is sent, the offer would remain open until the server restarts and sends the post.

There are cases where an offeree sends a rejection or acceptance, and then almost immediately changes his or her mind and sends the opposite response. When he or she sends both acceptance and rejection almost simultaneously, which response is effective depends on which one is dispatched first. If he or she sends a rejection first, the acceptance is effective *only if* the offeror receives it first. If the acceptance is sent first, the acceptance is effective upon dispatch and the rejection does not undermine the acceptance. The offeror's receipt of acceptance, in the offeree's simultaneous transmission of conflicting responses makes the acceptance valid upon receipt despite the fact that the rejection might have been sent first in actuality.

The Payoff: Consideration

The payment of money, action, or inaction, called "consideration," completes the trilogy of the requirements for forming a contract. Where an offeror makes an offer, giving an offeree the power to accept, and he or she does validly accept, either by acting on the offer or promising to act, the parties must complete the contract by providing consideration. A promise is supported by consideration if the promisee gives up something of value in response to the contract (incurs detriment), or if a promise is given as part of a bargain in exchange for something of value or for agreeing not to undertake some kind of action (and there is an exchange) (*Arledge v. Gulf Oil Corp.*, *Martin v. Federal Life Ins. Co.*).

Instances where money is exchanged or actions are undertaken or forborne in relation to an offer do not always count as consideration. For instance, a gift does not constitute consideration. And a promise to do something that the offeree has already done (preexisting duty) or would have done without the offer cannot constitute consideration. Consideration requires an element of detriment to the parties of a contract in order to create a bargain. When a party acts or provides a gift that gives him or her altruistic pleasure, no detriment exists; therefore, there can be no bargain. For example, a manager of a writing team is responsible for purchasing new computer equipment. She would feel good if she could help to support a friend who is starting a new computer sales business, so she casually tells him that she will help him get started by purchasing equipment from him. Neither mention price, model, delivery dates, or any other arrangements for purchasing the computer equipment. When the time arrives to make her purchases, she finds that she can save a substantial amount of money by purchasing from another supplier. Her friend cannot claim breach of contract because her altruistic motivation to buy from him did not create a contract. But consideration can be provided in the form of a promise to give a gift. Then the promise, not the gift, is the consideration. In this case, the promisor cannot decide not to give the gift.

Even though nominal consideration can make a contract invalid on the basis of sham, in general, consideration can be any amount of money, time, effort, or forbearance, as long as there is enough consideration to constitute a detriment, whether economic or not, and it is worth enough to the parties to the contract to be the basis of a bargain. Courts generally prefer to declare contracts valid rather than void them for lack of consideration. They often will declare a contract valid based on "implied promise" when an explicit promise does not exist. For example, a technical communicator asks a print shop and delivery service to print and distribute brochures for her client. The print shop's duty to mail, hand deliver, or otherwise distribute the brochures is implied, even though it is not stated explicitly in the contract.

There are a few cases where a promise is binding without consideration. Most states enforce promises to pay past debts, such as in the case of bankruptcy or debts that are beyond the time of legal validity (statute of limitations), which would void the debtor's liability. In addition, a promise to pay for benefits already received is especially enforceable without a promise. These cases most often arise when emergency care or aid of some form is provided in response to a request.

Meeting Your Contract Commitments and Avoiding Common Contract Problems

Once you begin to understand how the three contract elements—offer, acceptance, and consideration—are satisfied, you can look more deeply at some of the other, often even more complex issues involving contracts. Courts may find that the contracts that parties intended to create are invalid. The parties to a contract may also find that there are areas that lack clarity, in which case they may rely on courts to interpret them. Because these contract problems and their treatments are many and varied, they are organized here under a number of different headings.

Indefiniteness

One of the areas that can cause problems between parties to a contract (or potential contract) is indefiniteness. Under the Uniform Commercial Code (UCC) that controls commercial business contracts in most every state, if an agreement is indefinite there is no contract, but if the court finds that the parties intended to contract and the terms they agreed on are reasonable, the parties can treat the potential contract as a contract and can supply the missing terms. In other than UCC cases, the courts look to the test of reasonableness to decide whether a contract exists. In either case, UCC or non-UCC, if the court finds that the agreement is too indefinite, it will decide that there is no binding contract. A court can, however, allow the parties to include a contract clause that they agree to complete at a later point in time. Then the court usually supplies the terms based on the rule of reasonableness in price, time, or other category of indefiniteness. In the case of part performance, even if the agreement is too indefinite to enforce at the time it is made, partial performance on the contract can cure the indefiniteness. For example, a graphic designer agreed to provide a technical communicator with drawings to include in a report, but left the deadline indefinite. The technical communicator demands a June 20 deadline for the graphics because the overall report deadline is July 6.The graphic designer refuses and proposes an August 15 deadline instead. If the court finds that the graphic designer has a reasonable capability for producing the graphics by June 20, it will likely enforce the contract and resolve the indefiniteness.

Misunderstanding

Another problem area in contract law is in misunderstanding. When parties to a contract have misunderstood what they agreed upon, the court can find that they failed to understand each other's intentions and had no "meeting of the minds" and therefore, no contract. The parties to a contract have to have intended to agree, and the contract must reflect what they agreed upon. If the parties have different subjective beliefs about the terms of the contract, if a misunderstanding of the contract term is a material one that substantially affects the basis of the agreement, or if neither party knows or has any reason to know that there is a contract, the court will find that no contract exists. When a party knows or should know that the other party understands a term differently, then the contract is formed based on the understanding of the innocent party. The support for this kind of construction of the contract is based on the principle that one party should not be able to take advantage of the other. But if an offeree does not understand an offer and his or her negligence causes the lack of understanding, then the offeree is nevertheless bound by the terms of the contract. The fault for misunderstanding lies with the offeree in this case, so he or she should bear the burden of the fault. Conversely, where an offeror misrepresents a term in a contract and knows it, then the contract will be upheld on the terms understood by the offeree rather than the offeror. For example, a print company sends its new representative to sell print services to a communication company manager. He contracts with the new salesperson for stock paper, meaning the cheaper paper that the print company keeps in stock. But the new salesperson misunderstands, overcome by the zeal of knowing that stock can mean card stock, and the company prints the manager's order on more expensive card stock instead. If the manager knew that the new employee did not understand his order of cheaper stock, the print company could still enforce the contract for the more expensive paper.

Mistake

There are times when a contract might be formed on the basis of a mistake. Generally speaking, the legal definition of *mistake* is a belief that is not consistent with the facts. A mutual mistake is a case where both parties share the same mistaken belief, while a unilateral mistake is where only one of the parties has the mistaken belief. The parties to a contract may be either mutually or unilaterally mistaken about an issue of law where there is a misunderstanding about a legal principle. In any case, a mistake does not apply to an erroneous belief about what will happen in the future, but has to be based on an existing fact.

A mutual mistake requires three elements for declaring a contract invalid: (1) the mistake must concern a basic assumption upon which the contract was made, (2) it must have a material effect on the agreed exchange, and (3) if one of the parties to the contract bargained for an element of the risk of mistake, a mistake could void the contract only if it adversely affected the other party. In all cases, determining who bears the risk is based on standards of reasonableness.

Examples of each of the three elements will help to clarify the mutual mistake concept. The classic example in the case of a mistake that concerns a basic assumption upon which the contract is based, is that of a buyer and seller who both believe they have contracted for the sale of a Stradivarius but find that the violin is not a Stradivarius after all. But in the case where the violin was assumed to be a Stradivarius and was a Guarnerius instead, worth the same amount, the court would have to decide whether the mistake was "material," that is, meaningful to the extent that it would change the basis of the contract. If the substituted violin was of average quality and value, then the mistake would certainly be material and would void the contract. In the third element that could void a contract based on mistake, if the buyer, thinking that the violin might have been a Stradivarius, was willing to take a risk of paying more than he would have otherwise, he could not win a claim that the contract must be void on the basis of mistake. The seller, on the other hand, could, because she did not assume the risk of the sale.

Unilateral Mistake

The modern view on unilateral mistake, where only one of the parties to the work was mistaken, is that it is relatively difficult for the mistaken party to void the contract. Generally, however, to void a contract based on unilateral mistake, the parties must show the same three elements as in mutual mistake; plus, a party must also show either that it would be unconscionable to enforce the contract or that the other party had reason to know that there was a mistake and that the mistake was his or her fault. Most cases of unilateral mistake occur in construction bids where there are bid errors.

As long as a party did not fail to read the contract, even if his or her negligence led to the mistake, there is a chance that a court will void the contract or provide some other form of remedy. A court can void or rescind a contract, treating it like it was never made, or may award damages to a party who relied on the agreement to his or her detriment. A court may also reform a contract, which means it may recreate the contract. Normally, courts will reform a contract when the parties contracted orally in a situation that required a written contract, and they have acted as if the contract existed.

Conditions in Contracts

A condition is a contingency on a certain event, or an event that must happen before a contracting party is obligated to perform on a contract. When parties agree to exchange their performances on a contract at the same time, they can agree to concurrent conditions—conditions that they have to satisfy at the same time. For example, a technical communicator may develop a digital instruction manual for a company and agree to deliver the product on a set date as long as the company manager pays him the contracted amount on that same date. An "express" condition requires the parties to explicitly agree to the exact nature of the condition. A "constructive" condition, in contrast, is a condition determined to be a duty and imposed by a court. An express condition, for example, would be a situation where a company manager would pay the technical communicator for his digital instruction manual only upon approval. A constructed condition could follow the same circumstances, but the court would decide that the company manager must approve the product, despite the lack of language to that effect in the contract.

The law distinguishes between a condition and a promise. If a condition of a contract is not met, then the party expecting the condition to be met will not be obligated to perform his or her part of the contract. But if the contract is based on a promise rather than a condition, the party awaiting performance can expect to win damages for lack of performance. For example, a computer technician promises to repair nonfunctioning computers on the condition that she is provided with proper notice that they are broken. The requirement of notice is a condition rather than a promise. If the computer user does not give notice, then he has not breached (broken) the contract, but he has also not met the condition that would produce the technician's duty to perform. In this case, the technician does not have to make repairs, and the computer user cannot be sued for lack of notice provision. The court looks to the parties' intent to determine whether they desired to contract based on a condition or on a promise.

When Courts Enforce Commitments Without Valid Contracts: Promissory Estoppel

Promissory estoppel is a doctrine that provides for fair legal outcomes. This doctrine applies when one party makes a promise that forseeably induces another party to rely on the promise and then act in accordance. Under the promissory estoppel doctrine, promises that cause another party to rely on them can be enforceable without consideration (*Drennan v. Star Paving Co.*). Promissory estoppel is intended to avoid unfairness. For instance, an engineering firm's supervisor promises a young employee that she will pay for his college education if he attends full time. She intends the gesture as a gift, but the student gives up another job and enrolls in classes: thus, he owes the university for his education.

Under promissory estoppel, the engineering supervisor would owe the student the cost of his lost job's value in addition to his first year of tuition, even though the promise was intended as a gift rather than a contract offer. The requirements of promissory estoppel are that the person promised (promisee) actually relies on the promise and that the reliance is reasonably foreseeable to the person making the promise (promisor). Other circumstances that can generate promissory estoppel are charitable subscriptions in which a subscriber promises in writing to make a charitable contribution, or gratuitous "bailments and agencies," like a promise to look after another person's property or to act as another's agent. For example, an employee, acting outside her normal scope of duty to her employer, agrees to buy insurance for her employer as a favor. She forgets to buy the insurance and some time later when an accident occurs, the employer learns that he has no insurance. Even though there was no standard contract obligation, the employee is responsible for the employer's loss because he relied on her promise to buy the insurance.

Actions That Satisfy or Break Contract Commitments

"Substantial performance" is a state in which a party has acted in accordance with his or her duties to perform on the contract. A party must perform substantially for the other party's duties in the contract to be due. If one of the parties fails to perform, but the problems with his or her performance can be cured in reasonable time, then the other party's duties are only suspended, but still required. If a problem with a party's performance is not curable, or if he or she simply fails to cure it, then the other party no longer has a duty to perform on the contract. A lack of performance is considered a "breach" of contract, that is, breaking the promises made and breaking the contract. Because conflicts in contract law are rarely clear and simply decided, a court will determine not only whether a contract is breached, but whether the breach is material—important to the purpose and goals of the contract. The more likely there is some form of deprivation of benefit, the more likely the breach is material. The greater the extent of partial performance, the less likely the breach is material. The greater the likelihood that the breaching party can cure the problem with his or her performance, the less likely the breach will be considered material. A delay is usually not the basis for breach of contract unless the delay causes material harm to the purpose and goals of the contract, or unless the contract includes a statement to that effect.

Getting Out of Your Contract Commitments

Legal Exceptions for Not Completing Contract Duties: Impossibility, Impracticability, and Frustration

You probably know from your own experience that there are times when parties to a contract cannot or should not complete their contracted responsibilities, even when they are willing. These are situations of legal "impossibility," "impracticability," and "frustration." "Impossibility" occurs when the intended result of the contract cannot be accomplished. For example, an engineering firm contracts with a technical communicator to provide a report on its new jet project. The engineering firm's jet project contract is canceled due to the negative environmental impact of testing before the technical communicator can begin work, making it impossible to perform the contracted duties. Impossibility can occur in three ways: the subject matter of the contract preceding is destroyed, the means to carry out contract duties fail, or either party to the contract dies or becomes incapacitated.

In order for a contract to be voided based on impossibility, the action that is impossible must be one that is essential for performing duties on the contract. For example, if in the scenario just described, the technical communicator had agreed to report on jet project 1 but it changed to jet project 2, instead of being prohibited altogether, unless this was an essential element of the contract, the technical communicator would still be obligated to report on the project and could substitute a report on jet project 2. It is possible to void a contract on the basis of impossibility even if the impossibility is created by a third party's inability to perform. So if the technical communicator's reporting on jet project 1 was an essential element of the contract and the third party support for the work on jet project 1 was canceled, this form of impossibility would be enough to void the contract.

"Impracticability" occurs when new events arise that make it senseless to perform on the contract as the parties desire. For instance, where parties have agreed that one will create an advertising campaign for the other's business, and the business goes bankrupt and closes before the advertising campaign was supposed to have been created, the parties face a legal impracticability for performing on the contract. The contract is not impossible to perform, but it is so impractical that it makes no sense to try. Like impossibility and frustration, use of impracticability to void a contract must be based on a change of circumstances that occurs after the agreement was created and not on circumstances that already existed. The great majority of contracts voided for impracticability arise because a change in circumstances led to extreme increases in costs for goods or services where there are fixed price contracts. Generally, courts hold that sellers assume the risk of prices rising where it is foreseeable that this could happen.

"Frustration" occurs when new developments make it impossible to satisfy the purpose of the contract. For example, a multimedia designer agrees to pay an exceptionally high price to be taken on a set date to tour George Lucas's Skywalker Ranch during the development of his latest Star Wars film. In the meantime, the project has been postponed and nobody will be working at Skywalker Ranch on the Star Wars film. It would still be possible to tour Skywalker Ranch, but the purpose of the contract was to see the latest Star Wars film in development. The multimedia designer would be able to avoid contract obligations based on the "frustration of purpose" argument.

The doctrines of frustration, impossibility, and impracticability do not apply if the parties made arrangements for taking on the risk of their not being able to complete the intended contract purpose. So in the preceding example, where a multimedia designer contracted to tour Skywalker ranch, if the designer had agreed to tour Skywalker Ranch knowing that a Star Wars movie might not be in development, the multimedia designer could not make a claim for frustration of purpose. Contracting parties can always include explicit terms for handling contingencies and the courts will honor those agreements. Generally, where a party's primary purpose in entering into a contract is frustrated, the contract is discharged. This is a different case from impossibility, where a party cannot perform the contract duties at all. Here, the party can perform, but it makes no sense to perform because the contract's purpose would not be met. A court decides on two factors: reasonably foreseeability and totality. If the frustration was reasonably foreseeable, a court will usually uphold the contract. If the totality of the contract purpose is not defeated by frustration, again, the court will enforce the contract.

Generally, when a court decides that a contract's purpose is frustrated or the contract duties are impossible or impractical to perform, it tries to adjust the outcome so that both contracting parties are left in as good a position as possible under the circumstances. Either party may be able to receive the value of what he or she provided the other party, sometimes even if the value lies in preparation to perform contract duties. Under "restitution," usually the party who provided a benefit to the other can obtain the value of what he or she has provided to the other party. Very rarely, a party can obtain damages based on "reliance," where he or she can be repaid for expenses or effort expended in preparation for performing contract duties.

Miscellaneous Excuses for Breaking a Contract (Defenses to Breach of Contract)

Parties to a contract come into conflict when one party fails to follow through on his or her contractual obligations and "breaches" the contract. You have seen the many ways to breach contracts and to avoid performing on contracts throughout much of this chapter. Once a contract is breached, the breaching party may be able to provide a defense that excuses his or her legal responsibility for not following through on contract duties. These include defenses of illegality, duress, misrepresentation, unconscionability, adhesion, and lack of capacity, explained by category in the following text.

Illegality

Parties to a contract are not responsible for fulfilling their contractual duties if they entered into an illegal contract. In fact, an illegal contract cannot be legally enforced because it was legally prohibited from existing in the first place. Common kinds of illegal contracts include gambling contracts, lending contracts that involve usury (charging unfairly high interest rates), contracts where one person illegally finances another's lawsuit, and contracts for performance of services without a required license or permit. Employment contracts that include overly broad covenants (agreements) not to compete are common cases for the illegality defense because parties are often unaware that this form of contract is illegal. "Noncompete covenants" provide agreements from employees that they will not leave their current employment and do the same kind of work in competition with the employer in another setting. Where the terms of the covenant are reasonable and are intended to protect trade secret or customer lists, courts support them. But noncompete covenants that are so broad that they make it difficult or impossible for a former employee to work to support himself or herself are illegal and provide a defense for breach of contract.

For instance, a technical communicator signs a noncompete agreement whose terms provided that he would not release the company's customer list and that he would not develop documents for other companies. The technical communicator, assuming he would always work with print material, signs the agreement without reservation. After several years of longterm development of his skills as a technical communicator working extensively with multimedia production, the technical communicator is laid off from the company. He then begins to work for a competitor in multimedia development. His former employer sues on the noncompete covenant with a claim that he gave away the customer list and violated the agreement not to develop documentation for another company, digital or otherwise. The issues regarding breach of contract might be divided in this case. Courts sometimes apply the "blue pencil rule" where they essentially edit a contract and enforce noncompete contracts up to a reasonable limit and uphold some points of a contract, but leave others. If the technical communicator gave away the customer list, his former employer might be able to sue for breach, but the noncompete covenant is so overbroad that the technical communicator would not be breaching the contract by creating digital documents for a new employer.

Generally, illegal contracts are not enforceable, although courts may provide damages if one of the parties is unaware that the contract is illegal or if only one party intended the illegality. And where a statute declaring some form of contract illegal is intended only to protect one party, a court may go ahead and enforce the contract if it would benefit the protected party. When a contract has already been partly or fully performed, the courts are more willing to enforce it. If they can divide a part of the contract that is legal from the part that is not, the courts may enforce just that portion based on the "doctrine of divisibility."

Duress

If a party to a would-be contract is forced in any way to enter an agreement and can show that he or she was unfairly coerced, then duress can defend against a breach of contract suit. Courts use a subjective standard to make this determination, but they look to the likelihood that one party coerced the other based on the relative power between the two parties. If the court finds that one party could have had the power to force the other, then it next decides whether that party did coerce the other.

Misrepresentation

You may have encountered a situation in your own experience where one person tried to entice another by misrepresenting some quality that formed the basis for an agreement. The classic example is of the used car dealer who represents a used car as better than it is in order to entice a buyer into a purchase contract. Generally, if a party misrepresents an issue prior to signing a contract and then sues for breach of contract if the contracting party does not follow through, if the defendant can show that the plaintiff misrepresented the subject of the contract, the court will most likely rescind the contract or award the defendant damages based on this defense. The proof of the misrepresentation defense requires only a justifiable reliance on a promise and that the misrepresentation is a fact and not an opinion. Courts do not require a showing of intention to misrepresent. In fact, even negligent or innocent misrepresentation can be the basis for a valid misrepresentation claim.

Unconscionability

Sometimes a court may find a contract or even a clause in a contract so unfair that it would be unconscionable, or grossly unfair, to enforce it. There is no set legal definition of unconscionability, but the general issue is of one-sidedness. A court must decide whether the unconscionability would make the whole of the contract unenforceable or whether it could sever the one clause from the contract and eliminate the unconscionable effect.

Adhesion Contracts

Another breach of contract defense that implies an element of unconscionability is a claim that a contract is one of "adhesion." Adhesion contracts are documents that contain nonbargaining clauses in fine print, that are unduly complicated, or that are exceptionally favorable to the agreement drafter. An adhesion contract is either a document or a clause within it that creates a duty of one of the parties to do what he or she has not had the ability to bargain for. If a court finds that a contract or a clause within it was not negotiated, it often refuses to enforce the contract, especially if there is a great disparity in bargaining power. The court is even more likely to find adhesion clauses where a party does not even realize that he or she is entering into a contract at all. Courts usually will not enforce adhesion contracts unless they can find that the contracting party knew the effect of signing and assented nevertheless.

For example, a graphic designer signs a contract to use a special piece of visual tracing equipment. The contract includes a clause in the form of an adhesion contract that indicates her agreement not to hold the equipment provider liable for any potential damage to the work she is creating or to herself in the process of using the equipment. A court can declare the agreement an adhesion contract to uphold a policy that a company should not be allowed to provide potentially unsafe equipment and avoid liability in the event that someone is harmed by using it.

Adhesion contracts are basically documents that force contractors into agreeing to what they would not otherwise agree to because they have no choice if they want to pursue the larger, more significant contract goal. Today, there is potential argument that academic employment contracts that require professors to give away all their rights to the intellectual products they create are adhesion contracts. Professors often sign blanket agreements given to all academic employees upon entrance into university employment and know that they have no means to negotiate the terms of the agreement. Their only recourse is to choose not to take the job, but since the situation is the same in most universities, professors' power to make choices is very limited. The result is that they often sign contracts knowing that they have no bargaining power because they know that this is the only way to secure positions in their line of work. Courts have not spoken to this issue, but it may be an interesting legal question for potential decision in the future.

Capacity

Parties to a contract also must have the legal and mental capacity to enter into contracts. If a party to a potential contract is mentally impaired or under age, the contract will be void. In land contracts, though, where an agent contracts for the benefit of an infant, the contract would be voidable rather than void because the infant could make a decision to ratify (make valid) the contract once he or she comes of age. In order to rescind a contract based on mental incapacity, a court would have to make a determination that a party to the contract was not legally capable of understanding the impact of his or her agreement at the time the contract was created. This is based on the concept that a contract must be formed from a "meeting of the minds." Both parties must know what they are contracting to do, must know the effect of their agreement, and must have the intent to make the bargain as it is represented by the contract. A person who is mentally incapable cannot successfully meet these conditions: thus, he or she cannot enter a contract.

How to Cancel a Contract

As long as a contract is executory (where neither party has performed), the parties can create a mutual agreement to rescind it (cancel it). Although mutual rescission does not have to be in writing, the parties would benefit by a written document to that effect in order to avoid potential lawsuits in the future. If either one of the parties has fully completed (performed) the duties required in the contract, then neither can rescind the contract. The nonperforming party would still owe the performing party the extent of his or her agreed consideration stated in the contract. Nevertheless, as this chapter has already pointed out, a contract can be unilaterally rescinded (canceled by one party) if a party is the victim of fraud, mistake, or breach of contract. Someone who rescinds a contract simply cancels or terminates it.

Violating Contract Agreements

When a party to a contract thinks that he or she will be unable or unlikely to fullfill his or her duties on the contract, indicating the likelihood of lack of action on the contract will act in the same way that a material breach does. But where a party indicates that he or she will refuse to follow through on the agreements made, the response is considered to be a pre-perfomance breach of contract, an "anticipatory repudiation." If a party whose duties involve some kind of economic action, such as payment, is insolvent, or if one of the parties is incapable of following through on the contract, the other party can stop his or her own contract performance. If nonperformance is almost certain, the performing party can cancel the contract, but where it is unclear whether a party will be able to perform, the performing party can only suspend his or her obligation to act in response to the contract. He or she can also demand assurances that the prospective breaching party will eventually perform. Without these assurances, the prospective breaching party will have repudiated (clearly indicated that he or she will not complete contracted duties) the contract. For example, a technical communicator contracts with an automotive company to edit a procedure manual for building an electronic car. Before he begins to edit the manual, he decides that has taken on too much work for the year and would rather take time off to go on vacation instead. He calls the project supervisor at the automotive company and tells her that he will not edit the manual after all. His notice that he plans not to perform on the contract acts as an anticipatory repudiation. At this point, the automotive company is no longer bound to the contract. But if the supervisor wants the technical communicator to eventually fulfill the editing duties, knowing that he is capable, she can demand that the technical communicator give her assurances that he will fulfill the contract upon his return from vacation. Otherwise, the technical communicator will have broken (breached) the contract and could be legally responsible for financial payment or other damage obligations that the automotive company may incur.

Where an anticipatory repudiation occurs when a party makes it clear that he or she cannot or will not satisfy his or her duty in a contract—which in most states, allows the victim to sue before the time for satisfying the duty arrives—a "repudiation" (as opposed to anticipatory repudiation)—which can be in the form of a statement or voluntary action—occurs *after* the time that the party was supposed to have completed his or her part of the contracted duties but did not. This gives the harmed party the right to sue immediately. A party can retract an anticipatory repudiation until an event occurs to make it final. If a repudiation occurs, the harmed party can lessen (mitigate) the damages resulting from the repudiation by securing an alternative contract. If the repudiator does not retract the repudiation, or broken contract, then the harmed party can still sue. If the party who has been repudiated does not owe any contractual duty because he or she has already supplied it or because it is not yet due, he or she has to wait until the other party's performance is due before suing in order to provide enough time for the repudiating party to complete performance on the contract. If the harmed party has been owed a contractual obligation for installment payments, he or she cannot sue for the future installments until they are due. For instance, a photographer gathering materials to include in a proposal for a technical communication company's brochure indicates that he cannot do the work he had promised to do because he is too tired to complete all the jobs he has contracted to do. In most states the technical communication company can immediately sue. If the photographer fails to complete his obligation by the contract deadline, the technical communication company can sue without hesitation, regardless of the state in which the repudiation occurred. But if the photographer agreed to provide 10 photographs at intervals of one per month, he would be obligated only to provide those which are due in current or past months.

Treating Contracts Created from Unclear Bases: Oral Contracts, Statute of Frauds, and Parol Evidence

Statute of Frauds

The statute of frauds limits the validity of oral contracts. Even though most contracts are valid, even if oral, the statute of frauds invalidates the following oral contracts: suretyships—which are agreements to take on the debts of others, marriage contracts, land contracts, contracts that cannot be performed within a year, and, under the Uniform Commercial Code, contracts for the sale of goods over $500. There are times when an oral contract can satisfy the statute of frauds, as when a written "memorandum" states the terms of the oral contract so that it can satisfy the need for a written document. And courts have upheld oral contracts when the actions promised were completed within a year (*Hopper v. Lennen & Mitchell, Inc.*)

The statute of frauds does not prohibit an oral joint decision to eliminate the contract, except in the case of the sale of goods, but oral modification of a contract is a different situation and must be treated like a new contract. In some cases, when either party relies on an oral modification, a court may enforce it despite the statute of frauds.

Parol Evidence

The "parol evidence rule" is another important limitation on contract operation. It applies only to documents that are integrations; an integration is the expression of parties' agreement that the basis for what is intended to be their contract is in its final form. There are two kinds of integrations: a partial integration, which is intended to be final but not to include all the details of the parties' agreement, and a total integration, which is final and includes everything the parties intended. The parol evidence rule dictates that no evidence of other agreements is allowed to contradict or supplement the contract, because the document is complete as it was intended. In a seeming incongruity, if an ancillary (external) writing is signed at the same time, then it is considered part of the contract document and is not subject to the parol evidence rule. This ensures that the contracting parties are able to contract as they intended.

Generally, the parol evidence rule is intended to prohibit parties from introducing new documents containing new terms or contract guidelines after the parties have formed a completed contract that reflects their intentions to contract when they created it. The rule does not prohibit the parties from creating subsequent agreements in writing, such as agreements to modify the contract. But these agreements are treated separately from the original contract and are not a part of the original. However, if the original contract contains a clause noting that no oral modifications are allowed, then a court will enforce the clause and prohibit subsequent oral modification agreements. For instance, a multimedia designer contracts with an advertising agency to produce CD ROMs to hand out as party favors at the firm's 10-year anniversary party. They contract to produce the CD ROMs but fail to include detail noting the label design on the product. The multimedia designer refuses to develop the label design, noting that this responsibility was not noted in the contract. The advertising agency manager introduces initialed notes indicating the details of their agreements regarding the label design. The notes would be considered enforceable as part of the contract only if they had been signed at the same time the contract was signed.

The parol evidence rule does not apply if a party introduces evidence to show illegality, fraud, mistake, or lack of consideration. In other words, if a party can introduce evidence to show that there is no valid contract, then courts will consider this new information in deciding contract validity, despite the rule. In rare cases, if a contract includes a disclaimer (a clause stating that no representations have been made), some courts will prevent a party from showing that the disclaimer is false. Also, if the contracting parties agree to a specific condition required for the contract to be enforced, but the condition is not included, courts will allow proof of the condition, despite the parol evidence rule.

The parol evidence rule does not apply to courts themselves. They may supply a missing contract term if it is apparent that the parties wanted to create a contract and there is a reasonable way for the court to decide what the missing term was. The kinds of terms that courts typically supply are the imposition of duty of good faith and the duty to continue business. Although very few, some courts decide that employment contracts include an implied term prohibiting termination in bad faith. For example, some courts will prohibit an employer from firing an employee in order to keep him or her from receiving retirement benefits.

What Might Result from a Lawsuit on Your Contract

There are two kinds of suits for a breach in contract law. One is a suit on the contract, where the parties created a legally sound contract and the defendant (the accused person or business organization) has failed to follow through on the duties noted in the contract terms. This is called a suit for breach of contract. The other kind is where a plaintiff can bring suit in "quasi contract." This form of suit is applicable where the plaintiff is not asking that the contract be enforced, but instead, asks for payment (damages) for the actual value of the duties that he or she expected would be performed irrespective of the terms of the contract. It is possible to win a case in quasi contract when the contract is unenforceably vague, illegal, the parties are discharged as a result of impossibility of performance or frustration (where the ability to follow through on the contract is thwarted), or the plaintiff has materially breached the contract—broken the contract in a significant way.

Whenever possible, the courts will provide a remedy in one of two forms: "specific performance" or an "injunction." When a court orders specific performance, it mandates that the promisor perform the acts that he or she bargained for when entering the contract. When a court orders an injunction, it mandates that the promisor refrain from doing some kind of act, in accordance to what he or she agreed within the terms of the contract. Equitable solutions (remedies) resolve parties' conflicts by attempting to make parties whole again, as if there were no conflict. So if a graphic designer agreed to create a digital image for an advertising company and then refused after entering into a contract, the advertising company might ask for "specific performance"—that the designer follow through on his or her contractual agreement. Following the common pattern, a court would be more likely to force the designer to pay for the cost of having another designer create what she had promised to design.

Even though courts apply equitable remedies where possible, there are three limitations on what a court will order in equity. A court will grant equitable relief only if damages are inadequate to protect the injured party. Some forms of equitable relief are not estimable with certainty, and sometimes damages in the form of money are no substitute for performance. Although the courts almost never order "specific performance"—the actual bargained-for duties of the contracted party for a personal services contract—where an employer can show that the services are unique or extraordinary and the result will not be likely to leave the contracted party without means to make a living, the courts sometimes provide for specific performance and demand that the contracted party perform under the terms of the contract. For example, a digital film company that hires a famous actor to perform a role in its newest film would not be satisfied with money damages when it is the special character of the actor and his or her performance that provides the basis for their bargain in an employment contract. But a court will not provide equitable relief if the contract terms are indefinite or if enforcing and supervising the contract are necessary. So if the actor's role is not specified and it is unclear what would be expected in the contract, the actor would be unlikely to be forced to follow through on the contract.

Compensations to Winning Parties in Contract Suits

When a court provides a monetary remedy (damages) rather than demanding that a defendant complete the bargained-for actions in the contract, it will apply one of three kinds of damages—expectation, reliance, or restitution. Expectation damages are the amount of the benefit that the party expected from the contract. The goal is to put the plaintiff in the position he or she would have been in had the defendant fulfilled his or her duties in the contract. Reliance damages are those that put the party in as good a position as he or she was before the contract was made. The court usually awards out-of-pocket expenses that a party paid when preparing to try to follow through on the contract duties. This relief is used mainly when it is impossible to measure how much money the plaintiff could have expected or when the plaintiff wins based on promissory estoppel theory, where a party to a contract relied on the promise to the extent that he or she acted in accordance with the contract. (See pp. 66–67.) Restitution is a form of damages given when the court forces the defendant to pay the plaintiff the amount of benefit that the defendant received from the plaintiff's following through on the terms of the contract. Courts apply this remedy when a nonbreaching plaintiff (one who has not broken the contract) has partially performed and the restitution amount is more than the contract, and when a breaching plaintiff (one who has broken the contract) has not substantially fulfilled his or her duties but is allowed to recover the benefit of what he or she provided to the defendant.

Courts' Limitations on Detriments Resulting from Broken Contracts: Foreseeability, Avoidable, Nominal, and Punitive Damages

There is a limitation on what courts will award for breach of contract. They will not award consequential damages unless the damages occur from natural circumstances, and any reasonable person could foresee the possibility that they could arise. If the damages are remote and unusual but the defendant had actual notice of the possibility of these happening, the courts usually award them.

In any case, if the plaintiff could have avoided responsibility for the damage to the other party, a court will not make the defendant responsible at a later date if the plaintiff fails to make an effort to avoid the damage. These "duty to mitigate" damages are based on the test of reasonableness. A party must take reasonable action to lessen the potential damage caused by not following through on the contract duties. But plaintiffs are not required to ruin their reputations or undertake more than reasonable debt or hardship to follow through on a contract.

When a plaintiff has a right to sue for a broken contract, but no harm is done, a plaintiff can gain nominal damages—a small sum that is fixed without regard to the amount of harm suffered. But a court may also award punitive damages, or a penalty, for a broken contract if the act of breaking the contract also falls into the category of a civil illegality that is established in law (a tort). For example, an employer contracts with a seller to provide a certain level of safety equipment for employees, and the seller, who was supposed to have supplied grade A–level equipment, breaches the contract by providing grade B, knowing that B would be inadequate to protect the employees. The seller acted negligently or fraudulently in supplying the lower level of protective equipment and could be responsible for punitive damages.

Involving Others in Your Agreements: Assignment and Delegation of Duties

Assignments are legal structures that allow one party to an agreement to contract with a third party to take over his or her rights and duties to the other party with whom he or she contracted. But a promise to transfer rights in a contract is not an assignment because there is no consideration (payment of money, time, or action) required for an assignment. An assignment concerns three parties—the assignor, assignee, and the obligor (the party who was originally obligated to the assignor). Many states require that the assignment be in writing, particularly when a third party will receive payment from the original contracting party. In these cases, the assignee signs a document called a "security interest."

Gratuitous assignments are made in the nature of a gift, even though they're usually enforceable as valid contracts. They are irrevocable when they are supported by a written document, if the assignee relies on the assignment to the extent that he puts himself or herself into worse condition than before (relies to his detriment), or if the obligor pays the assignee for his or her performance.

Generally speaking, all contract rights are assignable. But where an assignment materially alters the obligor's duty (significantly changes it) in the case of a personal services contract, where there is a special relationship of trust, or if the assignee has some type of special ability or uniqueness, it is not assignable. A duty is also not assignable where it materially varies the risk in a contract, like that in an insurance contract. If the assignment changes the likelihood that the obligor's duty will be satisfied, an assignment cannot be made. In addition, if the contract terms include a clause prohibiting assignment, the courts will also prohibit the assignment. But the assignment will still be upheld as valid if the assignor fully performed the responsibilities contracted for. For example, a computer company hires a writer to create a brochure advertising its products. The writer realizes that she has so much work to do that she cannot complete the brochure, so assigns the contract to her friend, another writer. If the friend is not capable of producing the same quality of brochure that the computer company contracted for, or the contract contained a clause prohibiting assignment, the contract cannot be assigned. But if the friend fully completed the brochure to the same level of quality that the writer would have provided, the court is likely to uphold the assignment.

In addition, the right to sue for damages on breach of contract is assignable. If a clause in the contract states that contract duties cannot be assigned, then only the assignment of contract (delegation of duties) will be enforced and not the assignment of benefits from the contract. The assignment that violates an antiassignment clause does not make the assignment ineffective, but gives the obligor a right to sue for damages resulting from the breach. This determination will hold true unless a court applies rules of construction to see that the clear intent was to limit the contract from being assigned.

Third Party Beneficiaries

Sometimes parties will contract for the benefit of a third party, called a "third party beneficiary." For instance, if a technical communicator agrees to develop a brochure, but, rather than asking to be paid directly, asks that the fee be sent to the computer company that supplied the equipment for the job, the computer company is designated as the third party beneficiary. The computer company would be an "intended beneficiary" and have a right to sue for the fee due to it. If the promisee (the technical communicator) intends to give the benefit of the other contracting party's duties in the contract to the beneficiary, even in the case of a gift, the agreement will be enforceable in court. Contract defenses against the beneficiary are the same as those that could have been asserted against the promisee.

In contrast, an incidental beneficiary is one who benefits from a contract only by chance. Because the parties to a contract bargained for their contractual arrangement with no intent to provide benefit to a third party, he or she cannot sue for the benefit that would have been provided if the contract duties were completed. So where an engineering firm contracted to build a new plant in an area of town that could benefit from development, and a restaurant nearby would have benefited from the increase in potential lunch patrons, the restaurant owner would be an incidental beneficiary and could not sue for breach of contract if the engineering firm did not build the plant.

Ending Contract Responsibilities

Accord and Satisfaction
and Substituted Agreement

"Accord and satisfaction" allows parties to a contract to agree to substitute their duties in one contract for another. In an "executory accord," the parties agree that they will substitute the duties in the previous contract for a future performance that both parties agree on. In effect, they create a new contract by which they agree to cancel the previous contract. Executory accords are enforceable the way other contracts are but operate slightly differently, because if one party fails to complete his or her duties, the nonbreaching party can sue on the original contract rather than on the executory accord. A "substituted agreement" allows the parties to a contract to immediately discharge the previous contract and completely replace it with a new one. In this case, parties could sue on breach of contract, but only on the new substitute contract rather than on the original.

Accounts Stated and Releases

"Account stated" is a legal expression used to mean that everything that a party to a contract agreed to do has been done and that the effect of the contract has been satisfied and completed. An account stated gives a party to a contract the right to demand that the other party complete his or her duties in the contract. For instance, a multimedia designer completed a project for a client by the deadline that was set in the contract. The client owes the designer her fee for the work and the designer has waited a reasonable period of time for payment. She may now send the client a notice of "account stated" and demand to be paid immediately to complete the duties laid out in the contract.

You may know of cases where parties to a contract have worked together in a good relationship for some time and then one of the parties experiences financial failure. The solvent party, knowing that it will be difficult to be paid under these circumstances, and, based on a long-term relationship of good feelings toward the insolvent party, may decide to release the bankrupt party from his or her duty to perform on the contract. This can only be accomplished where the contract is executory on only one side, meaning that the party who decides to release the other has fully performed his or her own duty prescribed by the contract.

Discussion Questions

1. Can you list some examples of contracts that at first glance may not seem like contracts? Examples of seeming contracts that are in actuality not contracts?
2. A and B express a desire to share an apartment together. Is this a contract? If so, why? If not, why not?
3. Discuss the adhesion contracts that you have entered into in the past. What made them contracts of adhesion?
4. In your work or future work as a technical communicator, multimedia designer, digital video creator, graphic designer, or other creative communicator, what problems can you anticipate that could hamper your ability to join into clear contracts in the future?
5. What aspects of contracting will you pay careful attention to before entering into a contract in the future?
6. Are you in a contractual relationship now where not all aspects of your contract are clear? What are they and what is missing from the contract?
7. Are there times when breaching a contract can be beneficial?
8. Who is responsible for ensuring that the contracts you enter are clear and enforceable?

9. What kinds of clauses could you imagine in your own future work contracts that might be so unconscionable that they could be unenforceable?
10. Are there benefits at times to working without a contract? Why or why not? If so, what are those benefits?
11. To respond to the materials in Chapter 4, write a set of instructions for someone who works in your field or future field, taking the user through all the steps necessary for protecting himself or herself in the process of contracting with another party. You will want to draw from the material in the chapter to include helpful tips for ensuring (as much as possible) that he or she faces no legal problems arising from contracts in the future.

CHAPTER 5

Intellectual Property: Trademarks, Trade Secrets, Patents, and Copyrights

 You are likely to have read or heard about intellectual property issues, which is probably the most difficult area of law to understand. Nevertheless, intellectual property law is the most important for individuals who create intellectual products, as do technical communicators, multimedia developers, graphic designers, and other creative communicators. This chapter provides you with a general and relatively brief explanation of issues in intellectual property law. Its goal is to give readers who wish to gain a basic understanding of the law a starting point upon which they can build a broader understanding by reading other sources. Intellectual property law is sufficiently complex to require book-length treatment for explaining issues in more detail (as I provide in my book, *Controlling Voices*), but the material here sets the stage for further research in areas that individual readers may be interested in studying more deeply.

Intellectual property law is most important to creative communicators because it dictates what happens to every piece of work that you create. It decides how you may or may not benefit from your work, and who, in addition to you as the creator, has a right to benefit as well. It determines whether the person or business you work for or with is able to legally retain the right to control your work, and it forms the basis for the way in which you might create your work to protect your interests to your best advantage. Intellectual property law also is the constitutional base upon which our treatment of education, free speech, and personal development arises, which is important to creators both personally and in workplace settings; thus, this chapter will discuss these issues as well.

You will begin to develop a clear understanding of intellectual property law by first examining its foundation. Intellectual property law is based on a constitutional policy that supports knowledge creation and the development of new knowledge for the betterment of society. But in order to encourage learning and the knowledge-building process, intellectual property law provides incentives to creators by giving them rights to benefit from their work. To make free speech possible and to support a society that engenders egalitarian dialogue that forms the basis of policies that shape our nation's development and character, creators' rights are limited by public need to access information. The constitutional provision creates a balance between the needs of individual creators and the needs of the public at large. But its complex synthesis of historical treatment, interrelationship with other aspects of the law, and volatile interpretation and development resulting from the digitization of information makes intellectual property law extremely difficult to understand and to keep pace with in its continually adaptive forms.

This chapter is meant to provide you with a core framework for understanding the basic structure of intellectual property law. Because economic issues, rather than those regarding constitutional policy, are most often the focus for understanding the effects of intellectual property law on workplace issues, this chapter will treat more specifically the legal rather than the policy issues in intellectual property.

Intellectual property is so complex because it affects and is affected by so many other legal areas, such as contract, agency, and constitutional law. Let us begin by examining the different kinds of protection that the law provides for differing types of intellectual products. As communication creators, you will need to make choices based on which kinds of legal protection are applicable to the products that you create while being most likely to give you the kind of protection that you need. The most prevalently used categories of protection for intellectual products include:

- Trademarks, reputation, and goodwill
- Trade secret
- Patents
- Copyright

You may also consider protections such as mask works, character trademark and sponsorships, utility models, industrial design and design patents, invention and patents, and information data, but these are less commonly used and are not treated in this chapter.

Trademark	Trade Secret	Patent	Copyright
Symbolic icon Brand	Secret	Idea Mechanism System	Not idea, but expression Film Novels Plays Music CD Web material DVD

Figure 5.1 Legal protection for intellectual products

Sometimes intellectual products can be protected in only one of the category types listed, but at other times a creator may have a choice of how to protect a product and will need to decide which kind of protection accomplishes the goal most effectively. For that reason, you will find in this chapter a list of the protections and explanations of how they operate.

Trademarks, Reputation, and Goodwill

Trademarks, reputation, and goodwill are generally words or symbols used for distinguishing goods and services of one company or of one individual business entity from another. Service marks also fit in this category and are similar to trademarks, but they represent a service rather than a product. Trademarks provide a means to represent a company, product, or, in the case of service marks, a service. They can be in the form of text, graphics, or even sound. A trademark, in effect, is a tangible representation of the intangible quality of goodwill—the positive reputation that is attached to a company, product, or service. In a sense, trademark is the representative name for all three aspects of value, the mark itself, and the goodwill and reputation it represents. The trademark's value is in its representative character. An example of a trademark is Apple Computer's multicolored apple symbol with a bite in the side, which is quickly recognized as a symbol of the company. The fact that Apple product-users proudly display Apple stickers on their cars, carry Apple-branded computer bags, and wear Apple-logo T-shirts indicates the value of goodwill in the Apple brand. New developments in trademark law include the Anticybersquatting Consumer Protection Act, which authorizes a digital trademark protection that crosses digital boundaries and attempts to resolve domain name disputes globally (Nguyen).

To benefit from trademark protection, you must register your chosen mark with the US Patent and Trademark Office. If your registration is prior in time to all others, you gain an exclusive right to use the trademark to represent either a company or an individual. Once a trademark is properly registered, nobody but the trademark owner can use it to represent a company or a product. Registration protects against others who might want to benefit from a company's goodwill (represented by the trademark) for their own purposes or who might want to harm a company's goodwill by misusing the trademark to represent something not in character with the trademark holder's intentions. For example, consumers choose products based on the reputation represented by the trademark, so if a faulty product is illegally marketed with the trademark of a company that prides itself on producing a quality product, the reputation, and thus, goodwill, of the company that properly holds the trademark would be harmed.

Trademark registration provides notice that the mark is already "claimed" by a user and after five years of registration the trademark becomes "incontestable," that is, unchallengeable. This gives the prior registrant advantages in legal procedure in case there is a legal question about who maintains the right to use the trademark. Trademarks are federally registered to provide both notice and rights nationwide.

Choosing a trademark well is a careful process. You should complete a search to make sure that no other company or person is using a similar mark and that the trademark you choose is protectable. A mark cannot be registered if it already has a common usage. For example, a technical communicator developing product documentation and public relations material for an engineering company would not be able to choose and register a trademark in the name "camcorder," because the name applies to a class of items already in existence and produced by other companies.

Those who illegally misuse another's trademark may be legally responsible for damage payments based on harm to the rightful holder's goodwill. They are most often enjoined (prohibited) from using the mark, and often are ordered to pay damages as well. But at times, using another company's or person's trademark is protected and supported by the law. When trademarks are critiqued or parodied, this use can be protected under the US Constitution's first amendment clause as free speech. First amendment arguments and cases show that intellectual products of all kinds are subject to political criticism. For example, a court recently supported the Dutch rock group Aqua's parody of Mattel Toys' Barbie doll in their song "Barbie Girl" (*Mattel, Inc. v. MCA*

Records, Inc.). The song made fun of the Barbie "image" by creating sexual suggestions about the doll and the values it represents. The court supported Aqua's right to express an opinion through parody and upheld its use of the trademark for that purpose.

Trade Secret

Trade secret, another form of protection for intellectual products, provides the least protection of the potential choices you might make. Trade secret protections vary from state to state, but all carry the distinction of being the only protection that depends on the willingness of the secret holder to remain quiet and not divulge the secret. This makes trade secret a relatively weak form of protection, but one that may range from necessary to best choice of practice.

Trade secret requires two elements: (1) the need for a secret and (2) reasonable efforts to maintain secrecy. Fundamentally, a trade secret is a piece of information (that can be in the forms of a formula, pattern, device, or compilation of information) that allows one person or business organization to gain an advantage over another. Its power of advantage lies in the secret holder's willingness to maintain secrecy for his or her benefit or the benefit of the company that made development of the secret possible. Among the possible subjects of trade secret are the following:

- Customer lists
- Designs
- Instructional methods
- Manufacturing processes and product formulas
- Document-tracking processes.

Trade secret is often used in software development but can be used to protect any kind of product or idea creation. The Coca Cola Company may be most well known for keeping a trade secret. Its formula for Coke "Classic" has been a secret since the inception of the company. Trade secret provides a scope of protection that is based on a former secret holder's punishment for divulging the secret under these circumstances:

One who discloses or uses another's trade secret, without privilege to do so, is liable to the other if

a. he[she] discovered the secret by improper means, or
b. his[her] disclosure or use constitutes a breach of confidence reposed in him[her] by the other in disclosing the secret to him[her], or

    c. he[she] learned the secret from a third person with notice of the facts that it was secret and that the third person discovered it by improper means or that the third person's disclosure of it was otherwise a breach of his[her] duty to the other, or

    d. he[she] learned the secret without notice of the facts that it was a secret and that its disclosure was made to him[her] by mistake (Neitzke 26–27).

Since protecting trade secret and avoiding legal conflict is almost entirely dependent on trusting employees with access to secrets, there is no basis for allocating trade secrets by filing or registering them. The legal issues surrounding trade secret are based on deciding that a trade secret existed, that the secret holder was responsible for keeping it secret, that he or she did divulge the secret, and that disclosing the secret was a violation. There are six factors for deciding that there was, in fact, a trade secret:

- If information that is claimed to be trade secret is known outside the business, there can be no claim of trade secret.
- When many employees or others know the information that is claimed to be trade secret, the individual charged with disclosure is unlikely to be found liable.
- If an individual guarding the trade secret took extensive measures to protect the secret, yet it was still disclosed, he or she is less likely to be liable for trade secret violation.
- The extent of the secret's value can affect a decision about whether a secret holder is liable for trade secret violation. The more valuable the information, the more likely the holder is to be determined in violation.
- Likewise, the greater the amount of money or effort expended to develop the information, the more likely the holder is to be determined in violation.
- If the information making up the secret is easily and properly acquired or duplicated by outside parties, there is less likelihood that a secret holder will be determined liable for violating trade secret when the secret is disclosed.

In order for a trade secret to exist, the information must be unique from everyday knowledge. It must have a quality to make it unlikely that it could be developed casually, without support of a business or great individual effort in its creation. It has to be protected from anyone other than coemployees in order to be considered a trade secret, and it must be the kind of information that another company or individual would not be likely to develop individually. If a secret is reverse engineered or created outside the business claiming the secret, it is legitimately revealed, open to the public, and no claims of divulgence can be upheld. For example, a multimedia designer cannot claim a trade secret in VRML code that she pieced together from tips on Internet discussion lists.

More often than not, trade secret conflicts arise not with employees and employers, but with former employees and former employers. Despite contract terms that state clearly an employee's duty to maintain secrecy, dissatisfied employees or former employees often divulge secrets for revenge or simply out of lack of care in maintaining the secret. Employees or former employees who improperly disclose trade secrets may be legally responsible for the amount of damage that the disclosure causes the company, so they should be aware that their acts of disclosure may not go unnoticed or free of personal penalties.

Patents

Traditionally, patents have been used to protect tangible products such as mechanical devices. At one time when computer code was relatively new, patents sometimes applied to code as well, making intangible mechanisms patentable. Although computer code generally is not only not patentable today, but not copyrightable, the Patent and Trademark Office continues to patent intangible mechanisms such as management and business plans.

Words are not patentable, but processes, machines, composition of matter, and other procedurally based entities are. After a recent court decision in *J.E.M. AG Supply, Inc. v. Pioneer Hi-Bred Intern, Inc.*, even newly developed plant seeds are patentable. Patent protection is also applicable to ideas, in contrast to copyright (treated later), which only applies to expressions of ideas. Patents currently provide solid protection against reverse engineering and reproduction of the patented product, and have done so for the last 17 years. This enables patent owners to distribute their products without fear of violation.

Patents are difficult to obtain, in part because the patented product must be unique and its operations must be nonobvious. In addition, to satisfy legal restrictions, the product cannot have been described in a print publication within a year of filing or been sold earlier than a year before the date. The large numbers of similar products submitted for patents also increase the difficulty of meeting patent requirements. The process is difficult, expensive, and time-consuming, so you should carefully consider whether patent is the best protection for your intellectual product.

Once you receive a patent, you can exclude others from making, using, or selling the product you've patented. Thus, inventors are provided with incentives to spend time and money developing new ideas and new products that contribute to the country's bank of knowledge. But this system, although helpful, is not infallible. Difficulties arise when businesses, universities, and their competitors develop similar patentable software within a short time of each other. Patents are granted to the first application filed, so it can be risky to create products in competitive, fast-moving markets. If another product developer can show the "existence of prior art" (lack of novelty in the subject of proposed patent), your patent may be found to be invalid. For instance, an advertising copywriter loves using Post-it Notes while working on product development. But she decides that the product would be improved if it were available in multiple colors and not just yellow. A quick search of patent applications (and the local office supply store) allows her to see that this product already exists as a prior art, so she abandons the thought of patenting her new idea.

Copyright

Even though you might protect your intellectual creations in one of the categories already explained, you are much more likely to produce work that is more appropriately protected by copyright. Copyright is the most pervasive form of protection for intellectual products, especially for the kind of work—whether digital, graphic, textual, musical, visual, or aural—that communicates ideas through a unique form of expression. Court decisions about copyright are governed today by the 1976 Copyright Act, which is still controlling law. Notably, the 1976 copyright law includes explicit statements about what copyright does and does not protect. For instance, the law states that copyright protects expressions of ideas but not the ideas themselves. It does not protect facts, "any idea, procedure, process, system, method of operation, concept, principle, or discovery" (Sec. 102 (b)), even in authors' works. The law provides an explicit statement supporting public access to copyrighted materials, despite their copyright protection, in the fair use provision. It eliminates the requirement to register intellectual products in order to obtain copyright protection, and instead, provides that a work is automatically copyrighted if it is tangible, in a fixed state, and published (perceivable). The 1976 act also substantiates the legal fiction of corporate authorship through the work-for-hire provision.

You might think that a 1976 act would be old law, but the 1976 Copyright Act still applies and is controlling law today. The 1976 act was the first codified law to accommodate the technology of television, photocopying, and computers. Congress indicated its awareness of a change in our cultural understanding of intellectual property brought on by technological progress. Congress further indicated a mindfulness that intellectual products could no longer be treated like tangible property because they were no longer tangible. Instead, Congress's new statute reflected the policy goal in the constitutional provision, to support author/creators' limited monopoly in their intellectual products. In changing the characterization of intellectual products, Congress prohibited the use of common law proprietary argument—formerly used to establish authors' rights—and reaffirmed the statute's primary goal—to support learning.

As a technical communicator, multimedia producer, graphic designer, or other form of creative communicator, you will find that understanding the basic tenets of copyright will be helpful for making choices that allow you to take best advantage of legal protections and support for information access. Keep in mind that even though copyright is a distinct area of the law, it is applied in conjunction with other legal areas and is treated by legal concepts and arguments that are applicable in those areas. Often issues treating merger and reverse engineering, substantial similarity, contract law, agency law, plagiarism, tangibility and fixation, originality, and infringement arise in copyright cases and can influence case treatments and outcomes.

The bottom line for most authors and creators of intellectual products is that copyright provides protection against the threat of another's misuse of the work. Section 106, of the 1976 Copyright Act provides copyright holders with a limited monopoly in their work by authorizing them to control and use the work to:

- Reproduce the copyrighted expression
- Create derivative works based upon the copyrighted work (such as screenplay based on a novel)
- Distribute copies of the copyrighted work to the public by sale or other ownership transfer, or by rent, lease, or lending
- Perform the copyrighted work publicly, and
- Display the copyrighted work publicly

As you may remember reading, a creator does not have to file or register a work to obtain a copyright to it. As soon as the work is "'fixed' in a tangible form of expression (Title 17, U.S. Code), it is copyrighted. To be "fixed" means that the work is in a nonchanging form. Of course, this idea of "fixity" is difficult to pin down today, when arguably, it is rare that expressions are ever fixed because their digital character makes them changeable with the stroke of a key or replacement of a line of code. But most intellectual products are considered fixed if they can be recognized as whole, distinct from other products or from changed versions of themselves. When an intellectual product is in a tangible form of expression, it is simply perceivable. So the qualities of fixity and tangibility merge to some extent. To be recognized as a fixed entity requires that the intellectual product be perceivable and therefore, tangible. For example, a technical communicator's manual for developing corporate collaborative teams is both fixed and tangible when it can be read. As long as the concepts in the manual are only constructs in the technical communicator's mind, they are neither fixed nor tangible and therefore, not copyrighted.

Even though copyright holders are not required to register their copyrights, holders gain advantages when they do, so registration is well worth the time and the nominal fee required. Once a copyright is registered with the U.S. Patent and Trademark Office, the holder is provided with these benefits:

- Potential infringers are put on notice that the work is protected.
- The holder has the right to file suit on a copyright infringement claim.
- If the holder's infringement claim is successful, he or she can collect statutory damages.
- Successful infringement claims mandate recovery of attorneys' fees.

In contrast to what you have learned so far about copyrighting work, some works are not copyrightable at all. Materials prepared by federal government officers and individuals and employees of the government create intellectual products that are "public domain," in which the public has direct rights. State government officers and employees, as opposed to federal, have the same rights under copyright as other citizens do, and their work is protected under copyright to the same extent as that of others.

Copyright protection most commonly extends to literary works (like books, plays, or poetry), music, movies, and art, but can also protect videotapes, choreography, music videos, CDs, computer software, and other "original works of authorship" (Title 17, U.S. Code). Copyright applies both to literary works and to software products. Copyright contrasts to patent in that copyright protects both published and unpublished works, and patent only protects registered work.

Originality is also a legal requirement for copyrighting a work. An intellectual product does not need to be of good quality or real creative value, but only the result of an individual's or group of individuals' minds. However, originality alone is not enough. Facts, like scientific formulas, mathematical equations, historical theories, and other "verifiable" information can be original, in that one person may discover them, but facts are part of the public domain and cannot be copyrighted. What may seem to be in contrast to the way facts are treated, "derivative works"—creations that are based on other copyrighted works—can be newly copyrighted as original. Copyright holders have the right to create derivative works from their copyrighted work and prevent others from doing so, but in each new derivation a new copyright is established. A copyright holder may authorize a derivative to be created by a new developer, and the new creator would then maintain the copyright to the new work as long as the originality of the new contribution is "substantial" and only to the extent of the new contribution. For example, a computer interface designer might create a graphic to be used on a splash screen for an introduction to the digital work. But he may also create a derivative work from the same graphic by recreating it to function as a clickable icon that he uses in a catalog of all his works.

The statute authorizes other specific categories of derivative works, such as annotations, editorial revisions, elaborations, and compilations. Courts sometimes have difficulty in attributing copyrights in some compilations because they are based in facts, which are not copyrightable, although the compilations themselves are. Typical of compilation formats are telephone directories, catalogs, and dictionaries, which have no copyrightable merit in content but rather in the way the information is arranged. The layout and design that creators contribute to developing these works form the basis for the copyright in them. In effect, the creator is rewarded for the effort expended in creating the compilation.

Copyright Control

Copyright holders are granted rights to control their intellectual products through their legal definition as authors. They can license their copyrights, sell them, give them away, or transfer them in another form of arrangement as they wish. Copyright holders are called "authors" no matter what kind of work they create, and with the legal fiction of the "work-for-hire" doctrine, employers are considered authors for purposes of controlling products, even though they did not physically create the work. Authors can be composers, artists, software programmers, corporate and university entities, or writers. The constitutional phrasing that speaks about "authors and their writings" has been interpreted in the courts to apply to technological advances in creative works that the framers of the Constitution could not have considered. Even though the framers would not have been able to conceive of a multiuser sales network, such as the system behind Amazon.com's digital marketplace, their structure for dealing with issues of control over the intellectual products that drive it are still applicable. Copyright control extends for the life of the author plus 70 years, as of the recent passage of the Copyright Term Extension Act. The term for holding rights in copyright is 95 years beyond the publication date in the case of corporate authorship. The Supreme Court in its most recent ruling on intellectual property at this writing, in *Eldred v. Ashcroft*, supported arguments that the term extension to 70- and 95-year periods for holding copyrights is constitutional. Eric Eldred, the plaintiff, had created a digital access site for on-line books on the basis of their availability to the public after their copyrights expired. Eldred and his supporters argued that the new Copyright Term Extension Act extended the term of copyright protection into two lifetimes, which was such a long period of time that it made the time limitation built into the constitutional copyright provision meaningless. The ruling of the Supreme Court reflects its disagreement and the copyright term was upheld as constitutional. The Court cited the availability of fair use as a means for the public to continue its access to copyrighted material and bowed to the powers of Congress to make decisions about the law itself. Based on this decision, as a creator with an interest in protecting your work, you have the advantage of keeping control of your copyright for the benefit of your heirs or those to whom you give or transfer your control for a period of approximately two lifetimes. As a creator whose interest is in using other work from which to expand or develop your own, your access is now severely limited; those who wish to use film clips, photos of artwork or other visual representations, or music clips, among other works that were created by an individual after 1923, are prohibited from doing so unless their use is allowed by the fair use exceptions noted on page 106 in this chapter.

As someone who creates intellectual products, it will be helpful for you to know how you gain the legal status of author so that you will have a sense of whether your work may be protected under copyright and whether it is you, rather than someone else, who has the right to control the work. You may be considered an author under several circumstances:

- You independently author the work.
- You are an employer who pays an employee to author the work under the work-for-hire doctrine. (See the section on work-for-hire, on p. 97 for more detail.)

- You created a commissioned work for someone else, but because you contracted to do so, you still retain the copyright.
- You collaborated with another or other creators and you all hold rights in the intellectual product together. (See the section on joint work, p. 100 for more detail.)

Two categories of authorship that determine your extent of control and that can be hard to understand are work-for-hire and joint work. But the following explanations will give you a foundation for understanding how they work and should provide a framework for further reading and study.

Legal Fiction of Employer Authorship: Work for Hire

You may remember mention of the work-for-hire doctrine in Chapter 2. As noted there, a work-for-hire determination is dependent on agency law, but here you will learn that it comes from the language of the 1976 Copyright Act. Even though work-for-hire is specifically noted in the 1976 act, it has been broadly treated in legal cases since the 1909 Copyright Act. The work-for-hire doctrine creates a legal fiction that names the "author" of a work the employer or hiring party who contracted for it, rather than providing author status to the actual "author" or "creator." Since it is the "author," within the legal meaning of the term, who controls the copyrighted product, if a court decides that a hiring party is an employer and the creator is an employee, the hiring party obtains copyright control. You should keep in mind that the work-for-hire doctrine supports a *legal* rather than an *actual* attribution of authorship, so that in cases where your work is found to be a work-for-hire, you would not be awarded rights in the work that you actually "authored."

Remember that, broadly speaking, there are two major elements necessary for creating a work-for-hire product: (1) An author must be found to have produced the work as an employee, determined by a 13-element agency law test, and (2) he or she must have produced the work within the scope of employment and have no specifically contracted rights to the work. In the context of use in intellectual property law, the doctrine was designed to encourage new and creative work that would not be possible without the support of an employer, with the overall goal of advancing knowledge. Under the reasoning behind this doctrine, creators' incentives would come from the economic benefit provided by employers who could support new, creative, but expensive projects.

Before the days of telecommuting and extensive entrepreneurial work developed by individuals who have decided to profit by their

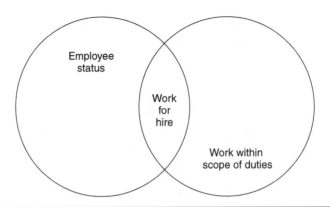

Figure 5.2 Work-for-hire elements

own independent efforts, determining employee status was much simpler. But now businesses are more commonly accepting the different kinds of work structures that technology use allows, where employees may work at home, in Internet cafes, on the road, on trains during commutes, and even while "on vacation" out of town or out of the country. So today, determining employee status has become more complex. Still, the courts make their determinations case by case by factoring in the circumstances provided in 13 elements for determining employee relationships under agency law. If you are producing work for someone else, you will need to know what effect your relationship will have on your rights to the work you produce. Likewise, if you hire someone else to produce work for you, this information will be just as useful.

So, are you or is the individual you hired an employee or an independent contractor? The context of your relationship, based on the 13 elements detailed in Chapter 2, pages 38–40, on agency will lead to a conclusion:

- If the hiring party had a right to control the manner and means for creating the product, the likelihood is increased that the work is for hire.
- The higher the level of skill required, the greater the likelihood that the creator is an independent contractor.
- Where a hiring party provides instruments and tools to create the intellectual product, the court will find support for determination of a work-for-hire.
- If the hired party worked at the hiring party's place of business rather than his or her own, this element will lead to a more likely finding of work-for-hire.

- The longer the duration of the relationship between the two parties, the greater the possibility of a work-for-hire.
- When the hiring party assigns additional projects to the hired party, he or she displays more control and is more likely working in the status of employer for purposes of work-for-hire.
- The more discretion the hired party has over when and how long to work, the more likely he or she is an independent contractor and can maintain control over the work.
- Hired parties who are paid by the hour, week, or month, rather than by the job, are more likely to be employees.
- When a hired party has a role in hiring and paying assistants, he or she may be legally determined to be a independent contractor.
- If the work created was something usually within the realm of the hiring party's business, this element could help make a showing that the work was not for hire.
- When a hiring party is not in business at all, it is harder for him or her to claim to be an employer.
- Hiring parties who pay benefits to hired parties are more likely than not to be employers.
- Hired parties who are taxed through the hiring party's business are more likely to be employees.

As noted before, determining work-for-hire status through the 13-element agency test is a cumulative process and relies on a court's understanding of the facts supporting all the elements in context. No one test from any one of these elements can lead to a determination. The Supreme Court considered all these factors in *CCNV v. Reid*. It concluded that because an artist, Reid, who had created a nativity scene depicting a homeless family huddling over a grate, had extensive control over his own workplace, hired assistants, and controlled the design and execution of the piece, he would be considered an independent contractor and could legally maintain control of his copyrighted work.

Notwithstanding the issues in work-for-hire, a creator may have developed a work that is considered a commissioned work, which is created specifically to be controlled by the party who commissioned the work. There need not be a work-for-hire relationship for the rights to the commissioned work to stand with the commissioner. There are eight kinds of commissioned works that are included in this category:

- Translations
- Compilations
- Parts of films or other audiovisuals
- Atlases
- "Consumables," such as standardized tests and answer sheets

- Instructional texts
- Supplementary works
- Contributions to collected works

Creators of commissioned works usually contract to produce a work on a one-time basis, and once the work is complete, the work relationship ends, or at least it ends for the purposes of this one product. The hiring party may ask the creator to develop a new product, but then the parties would contract for a newly commissioned work. For instance, a web designer creates a web site for a technical documentation firm. She is paid for the work and the firm employs its own web master to keep the site running and the content updated. The web designer's work in this instance is typical of a commissioned work. But if the web designer were paid by the technical documentation firm on an ongoing basis and acted as web master herself, the work would not be as likely to be considered commissioned.

If you are or become an employer, you would be wise to obtain an employee nondisclosure agreement from creators, particularly in instances when they contract to create commissioned works. This agreement can help to protect against claims that the works were created outside the creator/employee's scope of employment. But as a creator, you may also desire a contract that specifically states the nature of the relationship so that you can also be sure of what to expect from the relationship. (See Herrington, "Who Owns My Work? The State of Work for Hire for Academics in Technical Communication" and "Work for Hire for Non-academic Creators" for more detailed treatment of U.S. work-for-hire issues.)

Although a work-for-hire relationship cannot be changed by contract, parties may agree about who retains copyrights. As a creator, you may find it advantageous to relinquish your rights in an intellectual product for a higher fee or to open opportunities for future work. These, of course, are often more immediate benefits than what copyright retention might provide. If you are a potential employer, you could find it advantageous to pay more for work in order to benefit from it over the long term. Copyright holders may also license their work for a set period of time or for one use at a time, and thus, retain the copyright but bargain away the benefit in some ways.

Shared Authorship: Joint Work

Answering the question of who retains the rights to control copyrighted work becomes more complex when the creative product is a joint work. Basically, a joint work is a single work created by more than one author. Often joint works come about because the creators have different talents that they bring to a project to make it a whole. For instance, a group decides to make a digital film. One creator develops graphic designs, another, the sound track, and a third writes the script. The totality of the work is a complete, integrated whole, an entity in itself. In this instance, each creator shares the copyright in the work but would also retain the right to treat his or her piece of the work as desired. Therefore, each creator retains a copyright in his or her part of the work. More often than not, the parts of a work created through "joint authorship" are sufficiently merged to be inseparable so that the work has to be treated as one interdependent whole. But if you can and choose to sever different aspects of a work, each portion must be separately registered in copyright if it is to be considered protectable (*Raquel v. Education Management Corp.*).

Conflict in joint work usually develops when one or more of the creators of a joint work fail to agree that the group intended to create a joint work. Commonly, joint work producers develop their parts of a work separately and may not have even met or know who the other is, and in the case that the authors are employees creating a joint work for an employer, the creative product may be legally "authored" by a corporation or other business entity. The participants in joint work creation are not required to contribute equally to the project but, nonetheless, must provide some element of creative input to share the copyright. Regardless of their relative percentages of contribution to the joint work, all creators have an undivided interest in the work unless they establish a contract to the contrary.

Since circumstances for establishing joint work are sometimes unclear, courts look to intent to make determinations about whether creating a joint work was the goal of its developers. If a court finds that one of the creators had no intent to be party to a joint work but intended that his or her portion remain separate, the court will also find that there is no joint work. But there are times when the parts of a work become so intricately merged that its creators' different contributions cannot be separated from the whole. In these cases, courts look at all the issues within the intricate contextual situation to make decisions and may determine that a joint work exists despite intentions of the creators to the contrary.

You should carefully consider a desire to enter into a joint work project, because joint work authors are bound in an ongoing legal relationship after the work is complete. And in community property states such as Texas, California, Nevada, New Mexico, Arizona, and Louisiana, they are also legally bound to a relationship with the creators' spouses, who share their spouse-creators' legal interests in the work. Joint work creators must account to each other for profits received from use of the work, even though they can transfer their rights in the work without permission of the other. Creators you should keep this in mind if you enter into a joint work project, because you may want to create an agreement that all joint work creators must sign any kind of transfer contract.

Unauthorized Use of Copyrighted Materials: Infringement

Your benefit in controlling a copyright, provided by virtue of your status as an author/creator, author/employer, or joint author, gives you protection against unauthorized use, or "infringement," of the copyrighted work. An intellectual product user can infringe a copyrighted work in a number of ways. Where a copyright holder copies, performs, publishes, displays, or creates a derivative work from an intellectual product, he or she is acting within the range of rights provided by copyright in the work. But if a user who does not have a copyright in the work does the same, he or she is most often engaging in infringement of the work. (Note that personal use and the fair use exception, explained on pages 106–109, allow noncopyright holders to use materials under a range of contextual circumstances.) Not only does protection from infringement help eliminate economic exploitation of another's work, but as in the case of *Gilliam v. American Broadcasting Co.*, in which the Second Circuit found that the airing of a drastically cut version of Monty Python shows violated the integrity of the work, it serves to preserve creative intentions in the expression.

Users can infringe on copyright by making photocopies, scanning materials, performing, or reproducing works by some other mechanical means, or they may copy work on video or audio tape, CD, or displays. Proving infringement requires two steps. First, a court must establish that there is a copyright holder, and next, that a user has used the copyrighted work in a way that infringes upon it. It is relatively easy to prove copyright infringement when a user copies a work verbatim, but when the user has made so many changes to the work that the court finds it difficult to recognize the original portion of the work, infringement can be difficult to prove. Proving a work to be so similar to the original as to be considered a copyright infringement leads to a finding of "substantial similarity." An intellectual product may also be considered so common or usual that copying it might not be determined infringement of a protected expression. For instance, resume formats are very broadly used and copied without infringement. And a court would be hard pressed to find that common business forms such as formal reports, proposals, and memos would be infringing products. For instance, a multimedia designer copies a formal report format from a report he received from a technical communicator at an engineering company. The multimedia designer separated the elements, digitized them, and created an interactive DVD that users could implement as a ready form for report writing. The DVD is a financial hit and the technical communicator wants credit for her contribution to the product and claims copyright infringement. But the technical communicator is unlikely to recover damages because the formal report format is a common form and not subject to infringement.

Substantial Similarity

Courts most commonly find infringement in cases where it is relatively easy to show that a work is so similar to another that it had to be copied, which is called "substantial similarity." It is a requirement of infringement proof that the creator show that the intellectual product subject to the case is substantially similar to the work that he or she claims to have been infringed. If you read on, you will find later that the "clean room defense" can overcome proof of substantial similarity. To support an infringement claim, the claimant has to show that the defendant had access to the work and that it is substantially similar to the original work that is, enough similar to be mistaken for the same. Basically, claimants must show that the works are enough alike that copying was probable. The problem with proving substantial similarity arises when two works are sufficiently alike to support a presumption that one was copied, but where the copy includes substantial differences that can invalidate the presumption. For instance, two digital instruction manuals may use the same code and the same general organizational structure to convey the same instructional content. They may be substantially similar in their use of graphic illustrations that look much alike, such as blowup diagrams to convey ideas. But where one uses a different kind of arrangement and layout in providing the same content, it would be likely to defeat a claim of infringement.

The context in which a developer creates her own intellectual product determines whether there may be copyright violations based on substantial similarity. A variety of contextual circumstances can help to show that even though works may be similar, they are still not copyright violations. (See Lindey 12/5.) It is not uncommon that more than one creative product can share the same theme, and common themes tend to include common characteristics. For instance, screenplays depicting "buddy" themes often include two characters who are friends. One friend is smarter, more successful, a better athlete, or better looking than the other. The friends compete for the same goal but the more successful friend is kept by circumstance from pursuing the goal, so the less successful friend achieves the goal for both of them with the help of the more successful friend. This structure carries the characteristics common to the theme and can and has been depicted many times over within different settings. This example can also explain other circumstances that can indicate a lack of copyright violation. The example illustrates the use of stereotypes or stock characters and the same well-weathered plot. These kinds of entertainment are likely also to employ hackneyed ingredients, episodes, devices, symbols, and language.

Intellectual works' similarities may also arise when authors draw on the same cultural heritage or use the same traditional forms in developing their creative products. It may also be that convention requires set forms, like those in a three-act play or a memo. Competing creators may be influenced by the same sources or fashionable trends, or may depict the same cultural or historical event in a play. It may even be the case that similar works are created by mere coincidence. All of these circumstances could defeat an infringement claim. And in all cases, it is important to remember that copyright infringement only applies to a work's expression and not to its ideas.

Defenses Against Infringement

Some of the defenses against infringement claims in copyright include reverse engineering, merger, and the clean room defense (used when a creator develops a new intellectual product in isolation from the alleged infringed work). You might successfully use one of these defenses to show that even though your product is much or exactly like the product of the copyright holders who claim infringement, you have the right to benefit from the intellectual product you created. In the case of reverse engineering, for example, if you can prove that you recreated a product that already exists but through your own development efforts, you would not be legally responsible for infringement of the product on which you based the recreation. In a case that surprised many, the Court of Appeals upheld Microsoft Corporation's reverse engineered recreation of Apple Computer Inc.'s point and click user interface for its Windows application. The court pointed to the merger of the expression and idea of the interface, and in this case upheld Microsoft's use of the reverse engineered product (*Apple Computers, Inc. v. Microsoft Corp.*).

You will find that software development cases are common subjects of question in reverse engineering. Until the passage of the Digital Millennium Copyright Act (DMCA), there had been no inhibition against viewing the code in a software product and then, without copying the product itself, redeveloping it anew. This activity is questionably outlawed by the DMCA, whose effect is not yet completely clear (see pp. 110–111 in this chapter).

But up to now, the conflicts that the DMCA creates do not apply to reverse engineering of copyrighted products other than software. So, if a creator developed a new idea for a noncomputer game, for example, and expressed it through a series of descriptions and graphic explanations, another creator could study the original game to understand its operation and then reproduce it on his or her own without copyright violation. But the new game would have to be engineered rather than copied for the creator to avoid an infringement claim. Courts look to many factors to decide whether there is a valid reverse engineering defense. They examine the defendant's records to find a trail of memos or e-mails to substantiate that the work was done independently, and they look to see whether more than one creator's exposure to common influences could have led them to create similar products simultaneously.

The broadest and most accurate test for reverse engineering is the "clean room test," that shows that the product could not have been created without reverse engineering. The clean room defense allows a creator to show that he or she developed a product independently, in a "clean room." Some of the guidelines to make this showing are these:

- Occupants of the room have no access to the challenger's work or to any design or materials that would indicate how the work was created. Former employees are excluded from the clean room to avoid negative presumptions.
- The clean room is isolated from any communication between occupants and others who might have contact with the copyrighted work, design, or procedures for development. Even social contact is avoided during clean room activity. Physical location of the clean room away from social contacts also aids in maintenance of an untainted environment.
- The clean room maintains a credible monitor for all information, materials, and equipment that come in and out of the room. Often an independent third party is used for this purpose.
- Accurate and detailed records are kept regarding all information and materials that come in and out of the clean room and citations are noted when new items are produced in order to maintain records of their public nature.
- The monitor checks all communication that passes between occupants of the clean room and anyone else who has contact. All

telephone conversations are recorded and all correspondence, whether on paper, audio, or video tape, is dutifully monitored.
- All participants maintain complete records of day-to-day activities and progress made.
- The work created in the clean room should be seen and evaluated only at the end of all clean room activity (Galler 127–128).

The clean room defense requires extensive preparation and commitment to product development, even to the extent that some creators who work in a clean room environment are isolated from "the outside world" for long periods of time—living, eating, resting, and working in the clean room environment.

Another common and less strenuously prepared defense against infringement is merger. This defense applies when an idea and its expression become so intertwined that one is not separable from the other. Merger exists more often now than ever before, since digital products are so prevalent and digitization often leads to merger. For instance, informational CDs and even web sites often provide content by using graphic imagery mixed with audio bytes and even interactive functional elements to convey information. To delete any one of these elements would change the product to the extent that it would not function as intended. And where one designer contributes graphics that function in the interactive space and the other animates the graphics to make them operate, the product may be sufficiently merged to prohibit an infringement claim from the graphic designer when the animator benefits from the whole of the copyrighted product.

Personal Use and Fair Use

Other defenses against infringement claims support the overriding intent of the constitutional policy to sustain public access to learning for supporting free speech, education, and knowledge development. One is personal use and the other, noted within the copyright clause of the 1976 Copyright Act, is the fair use defense. (In *Basic Books, Inc. v. Kinko's Graphics Corporation,* however, the court sent a message that collections of copies compiled and sold by photocopy businesses will not be upheld as nonviolating under fair use.) Your role as a technical communicator, graphic designer, or other form of creative communicator often provides you with the legal status of author of intellectual products. You have seen how many legal means you have to choose from to protect your work. But you also play a role as a participant in national citizenship, and the intellectual property provision and the

Access to:
- All work for personal use
- Government works
- Public domain materials
- Works and uses exempted by fair use
- All other uses not prohibited by law

Figure 5.3 Constitutional support for intellectual property policy

1976 Copyright Act give you rights in that regard, as well. The fair use clause, pulled from the constitutional provision and explicitly codified in the 1976 Copyright Act, ensures that you and other citizens have access to intellectual information to enable you to maintain free speech rights and to participate in the development of knowledge that shapes what the country is and what it will be in the future. Your personal use and free speech rights protect your ability to use other work to comment on or to criticize cultural issues in our society.

Section 107 of the 1976 Act codifies supportive judicial law in the fair use doctrine, which supports the Constitution's goal and the primary goal of the 1976 statute, that promotion of learning is primary in importance:

§ 107. *Limitations on exclusive rights: Fair use*

Notwithstanding the provisions of section 106, the fair use of a copyrighted work, including such use by reproduction in copies or phonorecords or by any other means specified by that section, for purposes such as criticism, comment, news reporting, teaching (including multiple copies for classroom use), scholarship, or research, is not an infringement of copyright. In determining whether the use made of a work in any particular case is a fair use the factors to be considered shall include—

1. the purpose and character of the use, including whether such use is of a commercial nature or is for nonprofit educational purposes;
2. the nature of the work;
3. the amount and substantiality of the portion used in relation to the copyrighted work as a whole; and
4. the effect of the use upon the potential market for or value of the copyrighted work. (17 U.S. Code Sec. 107 (1978))

Fair use is intended to make information available to ensure an educated, free society. It allows intellectual products to be copied in order to support education, for purposes of criticism, news reporting, quotation, critical commentary, and parody. A critique may come in the form of criticism or illustration. Critical commentary is supported when it is used to clarify an author's observation or to support a point of critique or commentary. Critical comment, the basis of activity in free speech, can be founded on excerpts of another's work if it is used to show the subject of commentary and to make the commentary clear. Parodies are another form of critical comment that make visual or textual fun of the content of another work, and the substantially similar character of the parody to the work is the embodiment of the commentary. Therefore, this use of copyrighted work is allowed. In *Campbell v. Acuff-Rose Music, Inc.*, for instance, The singing group 2 Live Crew's parody of Roy Orbison's "Oh, Pretty Woman" song was upheld as a free speech expression of the banality of middle-class white attitudes regarding women. And in *Castle Rock Entertainment, Inc., v. Carol Publishing Group, Inc.*, the court decided that Carol Publishing's *Seinfeld Aptitude Test*, a humorous quiz book, was also noninfringing because it parodied the *Seinfeld* television show. But notice that the Supreme Court, in *Sony v. Universal Studios, Inc.*, has declared that all commercial uses begin with the presumption of unfairness. In addition, fair use in quotations has been limited to work that has been published for public venues; J. D. Salinger's unpublished letters were protected from copying, even under fair use exceptions (*Salinger v. Random House, Inc.*).

Libraries also may legally reproduce work to replace portions of works. They may provide summaries of works, as can others, either for personal or public purposes. Individuals may also use copies of works to support their cases in judicial proceedings or reports. Fair use even applies when a copyrighted work is copied fortuitously, as when a photographer or news reporter or other broadcaster is attempting to film another subject, but the work happens to be located at the same scene where the creator is filming.

As you have seen in other situations, courts look to a contextual set of circumstances to determine whether a defendant has a valid fair use claim. There is no set answer to any one fair use question. A court's decision about fair use is based on its interpretation regarding the situational context of use, purpose, duration of use, and substantiality of the work used.

One of the important contextual issues for deciding whether fair use is applicable is whether the user financially benefited from the copyrighted work and whether the use negatively affected the work's marketability. Fair use was never intended to support a user's ability to make financial gain, but only to support learning and free speech. Courts are generally not supportive of the fair use defense when the overall goal of using copyrighted material is to benefit economically. Likewise, they do not often support uses that harm the original copyright holder by lessening the marketability of the original intellectual product. Parody and criticism, however, by their very nature can harm marketability, but the overriding goal of supporting free speech overcomes the need to prevent economic harm. (See Herrington, "The Interdependency of Fair Use and the First Amendment," for more detail.)

New Developments in Intellectual Property in the Digital World

Digitization of information has changed the character of creative work to the extent that it has caused a rash of new applications of current law and the development of new laws in response. The ease with which digitized materials can be copied, in exact form, and quickly published or disseminated worldwide has caused fear that protected creative products will be misappropriated or mishandled. Corporate copyright holders, in particular, have lobbied for strict new treatments of intellectual products that sometimes prohibit use that was formerly upheld as legal just a few short years ago. Among the effects of digitization on intellectual products are the following.

Reproduction of Work in Digital Databases

Under the recent Supreme Court Ruling in *New York Times Co., Inc. v. Tasini*, the Court decided that freelance authors who previously published articles in periodicals would have to be compensated for republication of those works in web-based databases. This case supports an argument that republication of print work in electronic form creates a new form of publication that is newly copyrightable. Although the case decision led to eventual benefits provided to affected freelancers, today almost all freelancers are required to sign agreements to license their work for all forms of publication, print and digital included, which arguably is a loss of the benefit of choice. Although you are not bound in any way to license more use of your creative material than you wish to license, you should be aware of this trend in publication and know that you may be expected to follow it.

Deep Linking

"Deep linking" is another relatively new area of copyright law being considered now that text is digitized and the Web is a common publication medium. Deep linking provides a link to interior pages of a site and is usually created in order to bypass advertising or identifying information provided by the host home page. In both the first cases of deep linking in The *Washington Post v. Total News, Inc.*, and *Shetland*

Times v. Wills, the conflicts were settled out of court. But in *Ticketmaster Corp., et al. v. Tickets.Com, Inc.*, a California court decided that deep linking is not a copyright violation. These decisions may move web creators to use electronic means to prevent others from creating deep links, but the decisions point to a trend to uphold action as legal that allows the linking functions of the Web to be used to their fullest. It is likely that this area of law will be addressed again in the future.

Digital Millennium Copyright Act

The Digital Millennium Copyright Act (DMCA) is one of the furthest reaching limitations on digital information enacted into U.S. law. The greater the extent that your work is digital in nature, the greater the likelihood that you will be affected by the DMCA, which was enacted in 1999. Under the new law, generally supported by corporate software and entertainment developers and opposed by academics, scientists, and librarians, restrictions on both access and treatment of digital information are expanded.

The DMCA makes it a criminal act to create and/or distribute software that defeats encryption devices and that avoids antipiracy code built into much of the commercial software created today. It also makes it criminal to manufacture, sell, or give away devices used for illegally copying software. Even though the law arguably permits research on encryption or testing computer security systems, Princeton University Professor Edward Felten and Russian computer programmer Dimitry Sklyarov have been threatened and jailed for their work on encryption technology (Samuelson, Imfeld). Professor Felten and his research team took up a challenge by a group called the Secure Digital Music Initiative (SDMI) to find ways to defeat a digital watermarking system intended to protect music files from being accessed, copied, changed, or distributed. When Professor Felten's team of researchers from Princeton University, Rice University, and Xerox Corporation succeeded and tried to present their results at an academic conference, SDMI threatened them with a law suit based on the DMCA. The participants in the professors' research team withdrew their papers from the conference.

Dimitry Sklyarov carried out legal activities when he created a program in Russia that bypassed Adobe Inc.'s encryption measures for its e-Book software. When he came to the United States to attend a conference he was then arrested and charged with providing a device that could lead to copyright infringement. There was no charge that he had used the device or created it on American soil, but his arrest was based on the potential that he could use the device for infringement (*United States v. Dimitry Sklyarov*). This case is the furthest extension of copyright protection law to a form of digital expression and this, among the other broad-reaching dictates of the DMCA, has led to continually developing opposition to the act, based on claims of its unconstitutionality. In response to public pressure, Adobe eventually dropped the charges against Sklyarov but later sued ElcomSoft, the company Sklyarov worked for, charging it with violating the DMCA by creating a means to bypass its security settings for operating its e-book software. In December 2002, a jury found ElcomSoft not guilty, sending a signal that the DMCA may not be as enforceable as prosecutors expected (Berger).

In addition to the limitations noted above, the DMCA limits legal responsibility of nonprofits, universities, and Internet Service Providers (ISPs) for copyright infringement for simply transmitting material over the Internet, but simultaneously demands that they remove material from users' web sites if the material appears to be in copyright violation. (See Hayes for more detail on ISPs.) The DMCA requires "webcasters" to pay licensing fees to record companies, and although the language states that nothing in the law affects the rights enunciated under fair use, as previously noted, there is strong argument that fair use is not only restricted, but seriously at risk (Imfeld, Electronic Frontier Foundation).

Although the law is far from settled in cases that fall under the DMCA or are treated within the relatively new category of digital work, these narratives provide you with some insights into the development of law as it pertains to digital issues. You are likely to face the impact of new laws and their readings as you enter the workplace and, in addition to the laws that affect you within your own countries—as you note from the Sklyarov case above—the work you do can be impacted by the location from which you work. As the law develops in ways that reflect far-reaching global access made possible by digitized work, you will see that digitization brings out new questions: Whose law controls your digital content that extends beyond physical boundaries, and what is the effect of international trade law and human rights issues? (See Berman for extensive treatment of these issues.) In most cases you will not be touched directly by these broader issues within the realm of the work that you do. In the event that potential for your entry into these areas of unsettled law arises, it is best that you work with an attorney who can guide you in the specifically focused direction in which the law is moving at that time. For now, you are most likely to serve your needs by understanding the general operation of intellectual property law, while finding markers that guide you to areas where you may want to research more closely.

Discussion Questions

1. Which intellectual property law controls issues concerning the Internet?
2. Is the law different for digital materials and hard copy work?
3. What is necessary for creating a copyright interest?
4. What benefits are derived from filing a copyright interest?
5. How are your intellectual property benefits and duties affected by the types of communicative products that you create?
6. How do your purposes for use of intellectual products differ when you are using them for work and for personal use?
7. Should you register all your copyrights? Why or why not? What are the relative benefits and detriments?
8. How have you made use of your fair use rights in your educational career? In the workplace? In your home?
9. Describe two of the joint works you have created in the past. Will you knowingly participate in creating a joint work in the future?
10. How does context of use affect the legal status of intellectual products?
11. Your work requires or will require that you create many intellectual products over time. You will be faced with deciding how you want to treat your rights in the work you develop and you will need to be aware of the circumstances under which you will be able to retain the rights to your work. To make sure that you understand the impact of intellectual property law on your creative products, write an analytical memo that clarifies your interests. In the first part of the memo, make a list of the kinds of intellectual products you expect to create, and analyze the material in the chapter to determine which kind of intellectual property protection would be most appropriate for the kinds of work you have listed. Include benefits and detriments of each kind of protection and decide which kinds of protections can apply, as the law allows. In the second part of the memo, analyze the potential legal characterizations that could be applied to you in potential work settings. Note how, where, and in what kind of schedule and business arrangement you would like to work, and note whether you would be willing to purchase your own equipment and supplies and hire your own staff to do the work that you have in mind. Based on this analysis, you should know what to expect to do to protect your work, and what kind of employee or independent contractor relationship to expect to have with those who pay for the products you create.

CHAPTER 6

Synthesis

In the first part of this book, you read about agency and business organization relationships, contract law, and intellectual property law. I provided the information in segmented, isolated structures, treating one area of law after another to give you the clearest, simplest structure for understanding the issues that arise in these areas. Unfortunately, when legal problems arise in life, they are not segmented into structures that lend themselves to clear, objective understanding. Their interplay creates a synthetic and often complex relationship among the legal elements that arise in a given case situation. To help illustrate the synthetic (and more realistic) quality of legal issues in operation in real life, this chapter provides you with synthetic treatments of the issues explained in the chapters on agency and business relationships, contracts, and intellectual property law. This chapter should help you understand how different elements can work together within given contexts.

This chapter provides a variety of answers to the questions What is the importance of your legal characterization? What is the importance of the legal characterization of the product you are creating? What legal tools would you use to your best advantage in various contexts? and How does the kind of legal entity in which you are you working affect your rights and duties? The synthetic application of the law provided here demonstrates the effects of the different legal characterizations and explains possible legal outcomes. Because you will find the complexity of synthesis difficult to understand, even here, I provide you with isolated snapshots of only a few issues at a time. But you should note that each of the following snapshots builds progressively on your understanding of the preceding contextual construction and the issues explained by it, so you can build your understanding of the more complex integration of issues as you proceed from one snapshot example to the next.

Before you read the details of the snapshots it will be helpful to you to examine some general schema for how to break down the complex issues that arise in legal situations that involve multiple areas of law. Therefore, in this section, you'll find a series of diagrams meant to take you through increasingly more complex means of analyses.

You may analyze legal issues from a number of different perspectives, but here the snapshots put the technical communicator, multimedia creator, and/or creative developer at the base of the legal syntheses. This choice of perspective gives you a natural focus, since your questions and concerns will most likely arise first from how the law affects you as a creator and what it means for securing the products you create. The diagrams and explanations in this section will give you a foundation for building your understanding of the complex synthetic case scenarios provided in the next section.

Supportive Diagrammatic Schema

The first diagram represents the technical communicator, multimedia designer, or other form of creator within the box labeled "creator," along with his or her intellectual product, labeled "product." The intellectual product has a character of its own, and you must decide whether it is your own copyrighted work, "your copyright," or a "commissioned work." Remember that a commissioned work can fall within that which is covered by the work-for-hire doctrine that allows the employer to control the copyright, so determination is significant. Once you have decided whether the work is commissioned or your own copyrighted work, then you must choose how you want to protect the work. The different forms of protection are notated as "trade secret," "patent," "trademark," and "copyright." (See Figure 6.1.) Remember the significance of the choice of protection for your work, explained in Chapter 4

You should draw from the material provided in the previous chapters to understand the significance of these characterizations. But they will also be explained in detail in the following section, within the context of analysis of the complex case situations provided.

In the next visual framework, you can see that in addition to categorizing the creator's products, you must determine your own legal

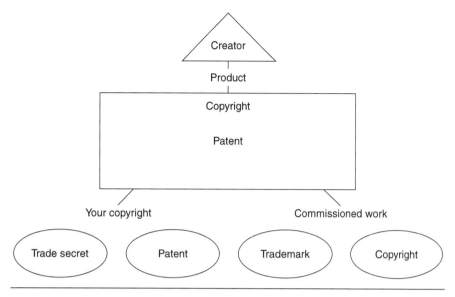

Figure 6.1   Multiple alternatives for characterizing intellectual products

character (as would other creators). In order to figure your legal standing, you first must decide whether you are an employee or an independent contractor, represented as "employee" and "independent contractor." As you know from your reading, this categorization is also important in determining who controls your copyrighted work under work for hire. (See Figure 6.2.)

Figure 6.3 illustrates the addition of contract complications, "contract," to the treatment both of creators' products and also of creators themselves, in their relationship to others. Notice that contracts can add complexity throughout many aspects of legal treatment. Chapter 3 provides you with information that would help you remember the potential effects of entering a contract and of possible breach of contract, which may be caused by your own or the other contracting party's actions or inactions.

You can see also in Figure 6.4, a visual representation of the material explained in Chapter 2, an illustration of how agency relationships between and among creators and those with whom they interact can affect not only the control of the creative products they develop but also all parties' legal liabilities. This added level of complexity is visually represented by potential effects that are just as varied in their intermingled forms as those of contracts. Agency relationships are represented as "agency."

Using the visual frameworks illustrating the relationship among elements of contract, copyright, and agency law, you can examine

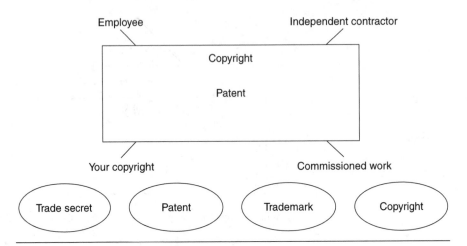

Figure 6.2 Additional variable in creator characterization

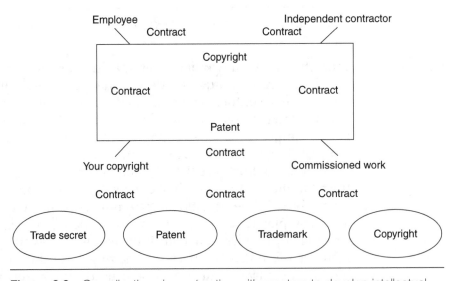

Figure 6.3 Complications in contracting with creators to develop intellectual products with multiple choices of characterization

case scenarios that depict the complex elements that intermingle in realistic contextual situations. Legal issues arise within the interrelationship of many circumstances surrounding law in the workplace but to understand how they can be treated under the law, you must separate them and isolate each circumstance so that you can analyze it within specifically focused issues of law. In some cases, you may also analyze the same element with several legal foci, for instance, understanding the effect of contract law as one focus, and intellectual

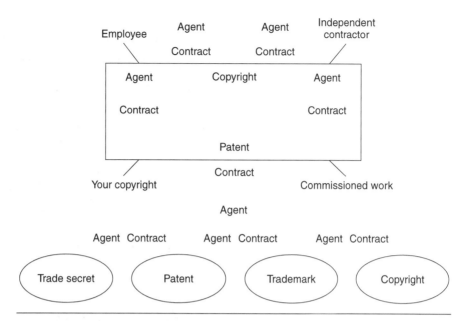

Figure 6.4   Agency relationships, control of creative products, and parties' legal liabilities

property law as the next. With that in mind, you will find here a set of case situations that provide syntheses of the areas of law discussed in earlier chapters, focused directly on the creator as the starting point of the different case scenarios.

Case Context (Synthetic Situation)

The case situation involves a creator, Spencer, who designs and develops a variety of intellectual products. Spencer works in an office in his home and uses equipment and supplies that he owns to create work for clients. This case involves Spencer's arrangement to create a digital atlas, what he calls his "Digatlas," for MapTech Corporation. On May 1, 2002, he meets Ben, a contact who tells him about MapTech's need for the atlas. They discuss Spencer's creation of the atlas for MapTech and they agree orally that Spencer will design and develop the atlas for MapTech's needs. They also agree that MapTech will only use the atlas for one year since the researchers at MapTech expect the geographic landscape of the world to change within the next year. They agree that the atlas should be completed by August 14, 2002, and that Spencer will be paid a set total sum, but will receive partial payment each month in incremental amounts until the work is completed.

Spencer goes home after meeting with Ben and begins work on the atlas. He buys materials to develop a prototype and starts work immediately. One morning, after a week of full workdays on the project, Spencer gets a call from his friend, Carmen, who is also a product developer, and Spencer tells her about his creation. Carmen suggests a way that Spencer can add flash to the Digatlas by developing a revolutionary digital mechanism to give it dimension through digital means, while also maintaining its ephemeral, transparent quality. Carmen and Spencer agree to work together to sell the new form of the product to the next company that will pay, while developing the work with the funds that Spencer receives from MapTech. Carmen agrees to share half the profits that Spencer receives from MapTech, to then work with him to produce and market duplicate products to future clients under the "Digatlas" label, and to share the profits and liabilities equally.

Spencer designs and codes the digital product while Carmen creates the enhancing mechanism that makes the atlas work in its multi-dimensional form. They are so enthused about the development that, working in Spencer's home office and using his equipment for two months, they finish developing the product ahead of schedule. It is at this point that Spencer finds that MapTech refuses to pay him further and also refuses to return the completed digital atlas to Spencer, claiming that the Digatlas caused eye damage to two MapTech employees who were so fascinated with the product that they stared too long at the light beams that display the digital information. He and Carmen also learn that Ben is the corporation president's brother-in-law and does not actually work for MapTech.

This case situation is complex and involves contract, agency, business organization, and intellectual property law, including work for hire. Analysis of this, like any other case situation, can only help explain the legal issues involved and potential case outcomes, but does not provide absolute answers for how such a case would be decided. With that in mind, you can follow the analysis to learn where to look for clues in deciding legal complexities such as those in the preceding scenario.

Case Situation Analysis

The first question to be decided is who, or more precisely, what, is Spencer? Spencer could be determined by law to be an independent contractor or an employee of MapTech. Spencer cannot control his legal status by contract, but must be determined to be one or the other by law. If Spencer is an independent contractor, he, and not MapTech, would retain the right to control the copyright to the digital atlas he created for the corporation. To make this determination, a court would look at the 13 agency law elements that determine employment status to see whether Spencer could be considered an employee. In this case, Spencer works at home, uses his own equipment, controls his own product development, "hires" his own help for creating the product, and is paid, although incrementally, for the product alone rather than for hourly work. By looking at those elements among the 13 elements of work for hire, a court would be likely to decide that Spencer was acting as an independent contractor. But given the circumstances of the context above, even though Spencer's status as an independent contractor indicates that he would hold the copyright to his work, the nature of the product he created can also impact whether a court will find it a work for hire.

Spencer's creative work is an atlas, which is one of the work products that falls within the category of creations called "commissioned works" and that are explicitly labeled as works to be controlled by the party who commissioned them, whether employer or otherwise. The court might possibly override the literal language of the statute and look at the broader range of circumstances to decide whether Spencer's work falls under the intended category for commissioned works, but there is no fine line to follow here to be sure that this would happen. The nature of the Digatlas may be considered different in that it is a unique digitized version of an atlas, requiring more creative effort than the more "factual" visual treatment of information in a standard hard copy atlas. A court might look to this element to declare that Spencer's digital atlas could not be characterized as an atlas as intended under the clause treating commissioned works, and could find that from that standpoint, the atlas could not be considered a work for hire based on this specific legal question.

A further potential complication in characterizing the nature of the atlas arises from Carmen's contribution. Carmen adds a "revolutionary digital mechanism" to Spencer's atlas, which changes the character of the digitized atlas sufficiently so that she and Spencer might have successfully applied for a patent in the product. If this were the case, then they would hold a patent that would prevent MapTech from claiming copyright control of the atlas.

Let's assume that the totality of the circumstances show that the Digatlas is not determined to be a work for hire. Let's also assume that it is protected by copyright rather than patent. If the Digatlas is not a work for hire, MapTech would not retain the copyright. But who would? The Digatlas was Spencer's idea. He pitched the idea to

Ben and was paid, at least partially, for it. But Carmen also contributed to the final product by providing the "revolutionary digital mechanism" that made it what it became. Both Carmen and Spencer contributed to the final product. The Digatlas was Spencer's idea, but under copyright, it is only the expression of the idea and not the idea itself that is protected. Spencer also contributed the content code for the product and the code necessary to digitize it. But Carmen supplied the "revolutionary digital mechanism" that became integrated into its final form. Her mechanism could not stand alone and Spencer's work, although it could exist outside the realm of the Digatlas, would not manifest in the Digatlas itself, but in a different product form. The result is that Spencer and Carmen's work cannot be severed from the Digatlas and have the Digatlas still remain intact.

A court would be likely to find that Spencer and Carmen created a joint work whose existence depended on both their contributions to the work. Ideally, they would have contracted with each other to determine their relative positions in relation to the work they created, but in the event that they did not and the court decided the joint work status, they would share equally in the benefits and responsibilities resultant from the work.

If, after the inclusion of the "revolutionary digital mechanism" in the coding for the Digatlas, Spencer and Carmen had submitted a patent on the product, their relationship to each other as joint creators would have been more clear, since they would have had to file the patent in both their names for both to be patent holders. Later, you will see that Spencer and Carmen could have created a contract arrangement to treat their product, giving them many options for sharing the benefits of their work.

You can see that just making the determinations that are visually represented in Figure 6.2 is difficult and requires consideration of many different complex legal issues that are further complicated when you move to the issues represented in Figure 6.2. Now consider how contract relationships complicate the case described, using Figure 6.3 as your road map.

The primary contract relationship represented in the case scenario is potentially between Spencer and MapTech. Ben and Spencer agreed that Spencer would produce the Digatlas for MapTech and that Spencer would be paid in incremental amounts for completed work. The contract was oral, but because it was not a sales or other form of contract whose validity would be prohibited by the statute of frauds, it could be considered valid. Spencer relied on the oral contract to his detriment, both economically and through his investment of time and effort in developing the product. He performed his duties laid out in the oral agreement by producing the Digatlas for MapTech. Thus, he supplied the agreed-upon "consideration," while MapTech supplied its consideration in the form of promised payments, at least until it refused to pay and retained the product.

At the point when Spencer completed the Digatlas and delivered it to MapTech, he performed his duty on the contract completely. MapTech was then obligated to complete its own duty and pay him in full for the work he completed. But MapTech refused to pay for the product, so Spencer now has a likely case for breach of contract. Although Spencer's claim to the payment he expects seems strong, MapTech might use several grounds to defend its refusal to honor the contract.

When Spencer met with Ben and they decided on the form of digital atlas that Spencer would create, the product description did not include Carmen's contribution of the "revolutionary digital mechanism" that makes up the Digatlas in the form that Spencer provided to MapTech. MapTech might defend its choice not to pay for the final product by arguing that the atlas does not conform to the product description agreed upon in the original bargain, especially since the new form, MapTech claims, causes eye damage and thus, would not be marketable. Spencer would strengthen his own claim to be paid if he could show evidence that MapTech representatives knew about the changes and agreed that he should continue to develop the Digatlas with the addition of Carmen's "revolutionary digital mechanism." But if a court found that MapTech's representatives were expecting the originally agreed-upon Digatlas, it might require Spencer to provide the agreed-upon form of digital atlas. A finding that Spencer provided a product that conformed to that required by the contract would be likely to render a decision that Spencer be paid in full for the product.

MapTech might also try to use the defense of "impracticability" to defend its refusal to pay Spencer (its breach of contract), based on the claim that the Digatlas causes eye damage and could not be marketed. But remember that the impracticability defense applies when, before performance, a party to a contract finds that acting on the contract terms is impracticable. This defense is not applicable once a party has completed work by the terms of the contract. This reasoning applies to the potential for voiding a contract based on illegality. Although it would be illegal to market a product, knowing that it could cause damage, the illegality defense applies when a contract itself is illegal, as, for instance, when a minor who cannot legally contract is party to the agreement. Given a situation where Spencer and MapTech worked together to support development of the Digatlas that includes the "revolutionary digital mechanism," and given that the product did cause eye damage, MapTech would still be bound to honor its contract with Spencer, even though a suit (or multiple suits among Spencer, MapTech, and Carmen) for legal responsibility on the damages, a separate legal issue, could follow. As you will soon read below, MapTech might have a valid defense to the contract based on mistake. In this case, it could claim that Spencer was mistaken in the belief that he could legally contract with MapTech through Ben.

MapTech's potentially more likely defense to a contract with Spencer is based on introduction of agency relationships, represented in Figure 6.4. Ben was not a member or employee of the MapTech Corporation, although he made Spencer believe that he represented it. In order for Ben to validly contract for MapTech he would have had to be an agent of the corporation. Without agency status connected to MapTech, the corporation could not control his actions, and neither could he represent the corporation and develop legal relationships between others and the corporation. But since there is no formal document required to create the agency relationship, Ben's lack of a formal agreement does not automatically dismiss the potential that he acted on behalf of MapTech.

The facts in this contextual situation do indicate a lack of agency relationship but also show that MapTech accepted at least the one act of Ben's "agency" in his contracting with Spencer to produce the atlas when it paid Spencer according to the terms of the agreement. Where the contract may have been void for lack of valid agency to contract, MapTech constructively accepted the contract terms by participating in the arrangement, knowing that Spencer would continue to rely on it and cause him to perform further duties in accordance. A court would be likely to find the corporation's actions an "expression of acceptance." In any event, if the contract were declared void, Spencer would be able to retrieve his Digatlas, since neither party would be allowed to be "unjustly enriched" by the relationship. Although courts rarely void contracts instead of upholding them where possible, if the court declared the contract void, it might provide Spencer with reliance damages for the time and money he spent preparing the product, which would put him back in as good a position as he was before he entered into what he thought was a contract.

Aside from the relationship between Spencer and MapTech through Ben's apparent agency, the questions regarding the actual agency relationship between MapTech and Ben is complex. As already noted, even though Ben was not MapTech's agent when he contracted with Spencer, he held himself out to be in an agency relationship with MapTech, and when MapTech responded to the contract by making partial payments, it legitimized his agency, at least to that extent. Payment on the contract provides evidence that the directors at MapTech knowingly allowed Spencer to believe that Ben was MapTech's agent, which provided him with implied authority to act as its agent. If MapTech refused to legitimize Ben's agency, it could have voided the contract with Spencer, but by allowing Spencer to continue to believe in Ben's agency, it became liable for Ben's actions on the corporation's behalf. Any arrangements that Ben had made after the time they created his implied authority would be legitimate on behalf of his relationship with Spencer. Ben's legitimacy might even extend to third parties, such as Carmen, who would have observed his agency relationship with MapTech in his dealings with Spencer and could have relied on his implied agency in her own dealings with MapTech through him. As an implied agent of the corporation, Ben's actions are attributed to the corporation, adding further legal complexity to the situation.

Another potential conflict over agency, in the realm of Spencer's potential agency relationship with MapTech, would arise in treating issues of legal responsibility (liability) connected with MapTech employees' eye damage, if it could be shown that the Digatlas was responsible. Since a company is legally responsible for the acts of its agents (employees), if a court determined that Spencer was an agent of MapTech, the corporation, rather than Spencer, would be liable for the harm. But since, as noted, Spencer is likely to be considered an independent contractor rather than an employee, he, rather than MapTech, could bear the burden of legal responsibility for damages. You may remember that independent contractors are not subject to a principal's (employer's) control, and are thus responsible for their own actions. The extent to which employees could claim damages as a result of their on-the-job injuries would be determined by the terms of their own employment contracts with MapTech. If the injuries occurred at the job site in the course of their duties as employees, they might have a claim against MapTech, independent of that against Spencer. A court would determine much of the extent of legal responsibility based on valid disclaimers that employees may have signed, along with state and federal regulation enforcement for safe workplace assurances. This material is not within the realm of what this textbook covers but would have to be considered in a case as complex as this.

As an individual acting alone as an independent contractor, Spencer would be personally responsible for any damages resulting from the product he created. But Spencer and Carmen created the Digatlas together, probably as a joint work. In fact, a court would most likely find that their agreement to split the profits and liabilities resulting from their product results in a general partnership between the two of them. As you know from reading Chapter 2 on agency and business entities, general partners share liabilities that result from the products they create, even if they have no written agreement to indicate this desire. And in a partnership, each partner would be personally, legally responsible for these liabilities. If Spencer fails to pay his share of damages, if a court finds them, Carmen must pay his in addition to her own share. In contrast, MapTech's liabilities are the responsibility of the corporation itself rather than its owners, managers, shareholders, and employees. Nobody would be personally liable for damages charged to MapTech.

Assuming that a court found that the Digatlas did not create employee eye damage after all, but that it was due to some other causation, the Digatlas could be a valuable product. Spencer and Carmen's partnership would be likely to face conflict with MapTech over who has access to the final product. MapTech's failure to pay Spencer for the product could result in a court requirement that it return it (among other remedies). In this case, Spencer and Carmen could continue to develop the product and sell it to other buyers in the future, probably more carefully contracting in writing with true agents of the companies they work with. But in the complex situation here, another interesting issue could arise. Noted earlier, the Digatlas is based on a "revolutionary digital mechanism." As you may remember, a mechanism is patentable if it is unique, nonobvious, and not described in print or sold within a year of filing for the patent. Spencer and Carmen might decide to protect the Digatlas under patent. Of course, if a court found that the Digatlas was sold to MapTech, that would invalidate the patent submission. But if a court declared the contract invalid for any of the reasons explained above, in the complex circumstances of this case, it would be in their best interest to patent the Digatlas.

Patent protection would ensure that MapTech could not reverse engineer Spencer and Carmen's product and recreate it on its own. The fact that MapTech had access to the product means that its employees would have been able to examine it and learn to develop it independently. Patent protection, although difficult to obtain, would give Spencer and Carmen protection against MapTech's use of their idea as well as the code that makes the product function. The content, in the form of factual information, would not be protectable in any form. But if Spencer and Carmen did apply for a patent, the case situation would still be complicated.

One important question a court would have to answer is whether MapTech bought the patent to the Digatlas or bought the Digatlas itself. If it bought the patent, of course, it, and not Spencer and Carmen, would have the right to reproduce the work and would be able to prevent them from reproducing it independently and selling it to other buyers. But if MapTech bought only the product itself, although this could prevent Spencer and Carmen from obtaining a patent, it would not prevent them from protecting their product under copyright law. Copyright provides weaker protection in some ways, but might provide the only feasible protection in this case.

Trade secret protection would not apply because MapTech made no arrangement to keep Spencer's work a secret and neither did Spencer make that arrangement with MapTech. Failing a situation that would allow patent protection, based on the "revolutionary digital mechanism" imbedded in the product, copyright would allow protection upon its creation in tangible, fixed form, once it was perceivable by a user. There is no need to file to obtain this protection, although, as you may remember, there are several advantages to filing with the copyright office.

If a court found that Spencer and Carmen held the copyright to the Digatlas, they could prohibit MapTech from copying the product and reselling it to other buyers. But they could not prevent MapTech from reverse engineering the product. However, MapTech would be hard-pressed to prove that the product was reverse engineered if they could not show that they developed it under clean room circumstances.

One more area for treatment in this complex setting is how to establish who is able to use the name "Digatlas" if either MapTech or Spencer and Carmen decide to file for a trademark in the term. Since a trademark is a word representing a company, its product, and the goodwill it embodies, it is important to the trademark holder that the product truly does carry the benefit of goodwill. If a court finds that the Digatlas does cause eye damage and the product is associated with the mark, neither MapTech nor Spencer and Carmen would benefit from holding the right to the trademark. But assuming either that the damage was not attributable to the Digatlas after all or that potential buyers have not yet heard the name, it would be of benefit to hold the trademark. If Spencer and Carmen registered the name before MapTech could, they would have a contestable claim to the name for five years. If MapTech could not show within that five years that the name represented it rather than Spencer and Carmen's general partnership, Spencer and Carmen would have an incontestable claim to the trademark "Digatlas."

As you can see, a synthetic view of some of the legal issues that might arise in any case situation renders truly complex and difficult questions. You should note that none of the issues is isolated from the other; each is intertwined and the direction a court decision might take can turn on one simple fact or action from one of the parties involved. The relationships among the parties to contracts, business arrangements, and working liaisons are integral to your understanding of the legal issues that can arise in real-life situations involving your development of creative communication.

The figures provided in this chapter should give you a starting point for analyzing and understanding problems you may encounter, or better yet, potential problems you can anticipate before they arise. This first part of this chapter provides a case situation that was focused on the central figure of the creator and, nearly as central, his work, since it was his work that created his relationships with all the players in this context. But you should note that you can analyze legal issues by placing other elements at the base of the analysis to simplify your desynthesis of the elements in a case in order to clarify your understanding of one issue at a time. For instance, MapTech would be interested in determining the agency relationship to Ben to decide whether to take legal action against Ben or against Spencer for introducing a potentially harmful product into the workplace, since the Digatlas was said to have caused eye damage. It might not consider the intellectual property issues involved at all until after deciding what its responsibility might be on the issue of liability (legal responsibility). It might prefer to try to support a claim that the Digatlas was not its intellectual product at all and thus, potentially avoid liability for any harm it might have caused. The next section provides you with examples of how you might examine the same legal issues just presented, but from different central perspectives or starting points.

Consider MapTech's interests and liabilities from its placement as the central focus of Figure 6.5. Just as in the first case, where your starting point was to understand Spencer's legal status as an independent contractor or employee, you must start at the same point when MapTech is the central focus. The facts in the contextual situation already presented indicated that MapTech is a corporation, which means that its members, shareholders, managers, and employees are protected from personal legal responsibility and share in profits and losses to the extent of their interests in the corporation. Through a legal fiction, a corporation is considered a "person" under the law.

After determining "who" MapTech is, you must decide where its pressing interests lie. MapTech has two central interests: control and benefit from the Digatlas and avoidance of legal responsibility for employee eye damage. It is retaining control over the Digatlas, presumably with the belief that the product has value. You can also presume

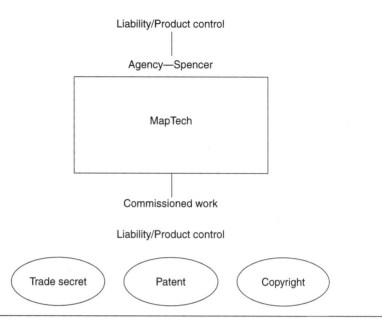

Figure 6.5   MapTech's interests, liabilities, and relationship to spencer

that MapTech has an interest in keeping the Digatlas since it refused to return it to Spencer. But since MapTech will also want to free itself of potential legal responsibility that might result from employees' suits regarding their eye damage, it might decide that its risk of legal responsibility is greater than the benefit it might receive from profits on the product. If you were in MapTech's position, after making this determination you would benefit by deciding on a series of ways to show that MapTech carries no legal responsibility for the Digatlas. To do this, you would need to explore issues in agency and contract law.

As you remember form the explanation given, if MapTech could show that Ben was not its agent, it could argue that he was not capable of contracting with Spencer on MapTech's behalf, and so the contract would be invalid; thus, the Digatlas would not be under MapTech's control, and the corporation would not be responsible for its assumedly damaging effects. Likewise, if MapTech argued that Ben was not its agent, then, he, and not MapTech, would be responsible for any ill effects of the Digatlas.

If MapTech decided that the case for legal responsibility in eye damage was weak and claimed copyright control over the Digatlas, it could choose to support its claim through contract and work for hire. Remember that MapTech is a corporation, but corporations carry the legal status of "person." With the status of "person," MapTech is legally capable of contracting with Spencer to produce the Digatlas. Of course, MapTech would have to be represented by a human agent to contract, but it could ratify the contract that was improperly made without Ben's agency, by accepting the terms after the fact. MapTech's payments to Spencer would provide evidence to show that this was the case.

In addition, MapTech might attempt a claim that Spencer was its employee working within the scope of his employment, making the Digatlas a work for hire and thus, its copyrighted work, but for reasons detailed in the contextual situation analysis, this claim would be weak. Without copyright control, MapTech would have to return the Digatlas to Spencer and would be prevented from creating derivative works from it as well.

You can see that placing different parties in the center of analytical application emphasizes different aspects of the law and different kinds of focused directions. Now consider Ben as the central figure. Again, it is helpful if you begin by deciding how to characterize the central Figure's legal status. You already know that Ben could not be classified as MapTech's agent in his initial dealings with Spencer. But MapTech's ratification of the contract, although it would not in itself ratify Ben's agency after the fact, could lead MapTech's managers to create agency status in him. As noted above, if Ben held himself out to be MapTech's agent with the corporation's knowledge and it did nothing to inhibit this, he could legally act as MapTech's agent, and the corporation would be responsible for his actions. If Ben wanted to be MapTech's agent, he might find this to his advantage, although it is unclear from the facts in the contextual situation how this would benefit him.

Where Ben's interests may be unclear, Carmen's are related to her time and skill investment in the potentially patentable Digatlas. As is consistent with the other situations already explained, you will be well served by beginning the analysis by determining Carmen's legal status. (See Figure 6.6.) Carmen might be Spencer's partner, contributor to a joint work, employee, or independent contractor. Most of the potential effects of this determination are already noted in the preceding analytical discussion, and are not repeated here. But it may be helpful to remember that once her legal status has been decided, then her choices for treating her contributions to the Digatlas are affected as well.

As you can surmise from the explanations given, you may want to proactively analyze a complex legal situation before entering into a legal relationship. By making one legal issue central rather than centering your focus on the creator or product, you may be able to clarify the

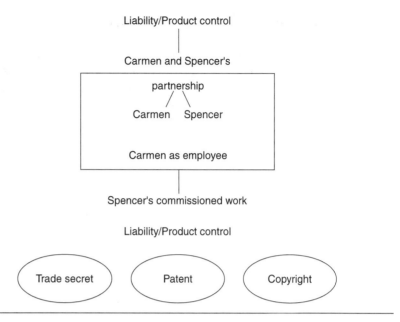

Figure 6.6   Impact of product control on liability

issues for yourself. For instance, when you begin to consider whether you want to create a product for a client, you might want to first decide what kind of professional relationship you want. Do you want to be an employee? A partner? An independent contractor? After deciding this first question, you would then need to consider the impact of your business relationship on the benefit you receive by producing the creative product. Without specific contracts to establish otherwise, if you contract as an employee, the employer would likely retain the copyright to your work. As a partner, you would share in the benefits of the work, as well as its costs and liabilities. The complexity of the legal issues involved in any given case situation can be defeated by careful desynthesis and analysis of each element separately.

It is extremely important that you become aware of as many of your workplace rules and parameters as possible before you agree to work within an organization or to produce creative work for someone else. As you can see, the complexity of even relatively straightforward business arrangements can be so overwhelming that in essence, it can hide the potential results of that relationship. The more carefully you consider your options for contracting with employers or clients, for protecting your creative products, and/or for establishing business organizations, the more confident you will be that you are achieving the highest benefits from your efforts.

Discussion Questions and Exercises

1. How would you construct a visual representation of the issues embedded in the situations described in this chapter?
2. In order to make your ideas for visual representation clear, map them and analyze each of the contexts provided in the chapter. Be prepared to explain your choices.

CHAPTER 7

Conclusion

Now that you have learned about the differences between law and ethics and about agency and business organizations, contract, and intellectual property law, you will need to consider how to apply that understanding to a complex contextual situation. You reviewed means to analyze how you might apply the laws in Chapter 6. In this chapter you will be able to apply your understanding to a complex case to test your understanding and abilities. Remember that an effective method for understanding complex contexts affected by distinct areas of law and ethics is to desynthesize the material to identify the elements of narrow issues before considering them in context. Answering the following questions may help guide your process of analysis of a complex series of issues in context:

- Who are the actors in this context? What is their legal characterization, and what impact does that characterization have on their rights and duties?
- To what business organizations do the actors belong, and how do their working relationships affect their legal rights and duties?
- What are the characterizations of the agreements (or lack of) between and among the actors in this contextual situation, and what is the effect of their agreements?
- What are each actor's rights and duties that result from their agreements?
- What is the legal characterization of the actions that the parties in the context have undertaken?

- How are the creative products developed by the actors characterized under law?
- Are the creative products treated in this contextual situation protected under the law, and if so, how?
- Who controls the creative products that are treated in this contextual situation and what is the extent of that control?
- What are the multiple potential options for future action on the part of each of the actors, and how might their choices affect their futures?

Keeping the material in the preceding chapters and these questions in mind, now consider the following case situation and examine the potential outcomes of the legal conflicts portrayed within it.

The Complexity of Law and Ethics Applied

As you have seen from the multiple examples in the preceding chapters, the distinction between law and ethics can sometimes be narrow, and that understanding the complexity of the different kinds of law as they intermingle can be difficult and tedious. Legal and ethical conflicts are built within distinct, yet complex, contextual situations in which there are no specific road markers pointing to direct answers for questions in the law. The following case situation contains multiple aspects of the law, provided in a mix of contextual details that require you to understand the materials from the previous chapters.

Case Situation

The TechComplicated Company was formed to create and sell documentation, communication, and web/digital development services to businesses that prefer to use contract work rather than employ their own multimedia staffs. TechComplicated contracted with a property company, LeaseUs Leases, to lease an inexpensive, though adequately apportioned and sized office space for its business location. When TechComplicated contracted for its 30-year lease, the space, located in Decotown, an older area of the city, was surrounded by interesting architecture but within a mix of poverty stricken residential areas, decaying office buildings, and scattered industrial facilities. Other companies like TechComplicated also began to locate their businesses in the area and after five years of rejuvenation, Decotown has become one of the most popular areas of the city, filled with coffee roasters, upscale restaurants, and renovated residential bungalow houses within walking distance of the life of the area.

TechComplicated was created as a limited partnership and is managed by its two full partners, Olaf and Gabrielle, who share responsibilities for project management, employee hiring and firing, and general company administration. TechComplicated's employees include a staff of ten writers, graphic designers, web designers, and other multimedia developers. It also contracts with additional creative communicators when a job requires someone with special skills or when the workload requires extra help.

Recently, Olaf and Gabrielle have been relying on long-time employee Meghan to help with some of the management duties of the company. Meghan often interviews long-time clients to determine their needs for progressive media jobs—needs such as web site updating and maintenance, hard copy brochure content update, redesign, and reprinting, screening telephone and walk-in client contacts, making office supply orders, and handling basic interoffice communication, such as office memo updates.

TechComplicated's largest client contract and longest-running client relationship is with Infoboom Corp. Infoboom is a fast-growing, medium-sized corporation that deals in all manner and measure of communication technology, from telecommunications to print media. Infoboom's assets include hard product and equipment as well as intangibles, such as goodwill. In addition to salary, it provides its employees with compensation in the form of stock benefits both in Infoboom and in other corporate holdings, including those of Endrun, Incorporated. Infoboom employees have been optioning their salaries to purchase Endrun stocks recently, based on media projections that Endrun will become the most powerful corporation in the country within the next three years and in the world within the next five.

You may already be considering the number of potential legal and ethical complications that may arise from the business relationships described. Embedded are issues of contract, agency, business entities, and intellectual property law—the legal areas described in the body of this book. The next paragraphs describe the companies' and employees' interactions. You should read on with the goal of discerning the legal and ethical issues that arise within their dealings.

TechComplicated has contracted with Infoboom to develop a new web site to launch Infoboom's new wireless network, YRless. Infoboom agrees to pay TechComplicated a set fee up front, a payment upon completion of the dummy site, and a final payment upon completion of the final product that includes content, providing that TechComplicated completes its products by set deadlines. A contract provision states that if TechComplicated fails to complete its work by the required deadline, Infoboom then has the option to continue with the agreement or cancel the contract. TechComplicated accepts Infoboom's first payment and begins work on the product. Because Infoboom's developing reputation will be reflected in the web site and because TechComplicated's managers know that their own ability to impress clients and increase their business presence depends on the quality of the web site, they decide to hire extra web developers and graphic designers with special skills to help with the YRless site development.

Meghan, TechComplicated's employee, contacts the candidates for the work and creates a list of five additional future employees, three who have worked with TechComplicated as independent contractors in the past. Olaf and Gabrielle approve of everyone on the list except Ester, who has worked for them in the past and produced inferior work. Meghan, having carelessly included Ester on her call list, begins to contact each of the five to offer them work. Meghan completes calls to Ester, Ikor, and Emp, who have worked as independent contractors for TechComplicated in the past. They agree to work on the project, and she sends each of them contracts specifying the work that each needs to complete. Olaf and Gabrielle then tell Meghan that the contracts sent to workers on the project must include a clause noting their agreement to relinquish any potential copyright control in their work. Meghan is embarrassed and worried about the mistake, particularly considering how anxious Olaf and Gabrielle are about the Infoboom contract, so she decides not to tell them what she has done. Besides, Meghan learns about the problem on a Friday and wants to spend the weekend relaxing rather than thinking about office problems.

Meghan gets some relief when she finds that she mislabeled the envelope with the contract for Emp and can resend a contract that includes the clause relinquishing copyright control. But she also finds that TechComplicated has received signed contracts from Ester and Ikor that do not mention the allocation of copyright control. Meghan makes the rest of the calls to the other workers, provides the appropriate contracts, and soon stops worrying about the problem. The work progresses well and without problems at stage one. Ester works from home, using a home computer to produce her work and send it electronically. In the chaos of product development, Olaf and Gabrielle fail to notice that Esther is contributing to the project because she sends her material from her home work site and because the other workers are handling most of the project development on the mock-up.

As in all aspects of life, all things cannot always go well, and soon TechComplicated is contacted by LeaseUs Leases. LeaseUs Leases has learned about TechComplicated's contract with Infoboom and, expecting great commercial success from both of them, demands that TechComplicated pay a large rent increase on its leased property. LeaseUs Leases argues that the property values have increased so substantially as a result of the development of Decotown that it would only be fair that TechComplicated pay an increase. In addition, it argues that other potential lessors have been clamoring to rent the space and offering much more money. LeaseUs Leases concludes with the complaint that the property valuation has increased so substantially that its tax increases are difficult to pay as a result and that TechComplicated should help bear the burden. TechComplicated firmly states that the business is responsible for exactly the price contracted in the 30-year lease, signed at a time when the property value was extremely low. Olaf and Gabrielle further argue that their participation in building the neighborhood has benefited LeaseUs Leases by increasing its investment in the property, thereby making it desirable to those who are clamoring to rent the building. They refuse to pay any more than the contracted rent price.

LeaseUs Leases' pressure is only the beginning of TechComplicated's problems. As the YRless team works frantically to produce a quality product and still meet the project deadline, they find many mistakes on Ester's portion of the Infoboom content area of the web site. When Olaf and Gabrielle finally track down the source of the problems, they realize that Ester has been working on the project and, in usual style, producing poor work. Meghan has to confess to her mistake in hiring Ester, and Olaf and Gabrielle call Ester to rescind the contract, claiming that she signed in bad faith, knowing that working from home would hide her participation in the project and allow her to continue to work with TechComplicated—despite her knowledge that Olaf and Gabrielle would not have hired her. But Olaf and Gabrielle are surprised when Ester retaliates by telling them that she has decided to maintain her copyright in the final product, based on her status as an independent contractor. Olaf and Gabrielle end the conversation and immediately start a careful review of all the contracts. To their dismay they realize that Ikor, in addition to Ester, has signed a contract with no clause relinquishing copyright control. Since Ikor has been working at the TechComplicated office with the rest of the team every day since the project began, they never questioned his employment status.

Olaf and Gabrielle consider firing Meghan but know that under the stress of the YRless project deadline, they could not spare the time and energy to hire and train someone new to take Meghan's place. They threaten to fire Meghan, but instead renegotiate her contract and decrease her pay. They also make it mandatory that Meghan get their review and written consent before acting on any decisions that affect the company.

While developing the content section of the YRless web site, the project team members find material that is unflattering to their client, Infoboom, and begin to suspect that Infoboom is working closely with Endrun to inflate Endrun's image by supplying employees with Endrun stock. The team members speculate that the information was accidentally included in the material that they were given for the web site. Olaf and Gabrielle, realizing the volatility of the information they have, command their team members to keep the information quiet until they decide what to do with it.

In the meantime, at Infoboom's office, rumors have begun to circulate that Endrun has been involved in shady business practices and that the information is about to be made public. Infoboom managers tell their employees who are Endrun stockholders nothing. The managers know that Endrun's failure would mean sure bankruptcy and decide to continue benefiting from Endrun employment as long as possible before the company fails. In order to avoid paying TechComplicated for the expensive work on the YRless project, and knowing that YRless is now unlikely to come to fruition, Infoboom's managers look for ways to renege on the contract. They decide that if they can stall long enough and hold back the necessary content for the web site, TechComplicated will not be able to create the final content for the site in time to meet the deadline and, as a result, will breach the contract, giving Infoboom the opportunity to back away and save money.

As it turns out, one of Infoboom's managers is a friend of LeaseUs Lease's owner, Owen, and one day over lunch he tells Owen about his worries regarding his future at Infoboom and asks for a job with LeaseUs Leases. Owen already knows about TechComplicated's relationship with Infoboom and immediately goes to TechComplicated's office to tell what he knows about Infoboom's expected financial problems in an attempt to pressure Olaf and Gabrielle to sign a new lease for a higher price. Owen tells them that if they sign the new lease, they can sublet the building if their business collapses as a result of their relationship with Infoboom, and that he will keep the information quiet long enough for them to complete their YRless web site and obtain the inevitable log of new clients as a result. If they refuse to sign the new lease, he will make the information public. He gives them two weeks to think about the proposal.

Olaf and Gabrielle are worried about LeaseUs Leases and about Infoboom's possible collapse, but more pressing is that they are unable to obtain the content material necessary to complete the YRless web site. Emp, knowing about the unflattering material that the project team uncovered, and watching Olaf and Gabrielle's anxiety over developments that they refuse to mention to the team, begins to think that TechComplicated may face an unstable future. His friend, Topit, who produces award-winning technical documents and digital materials, comes to him one day, excitedly telling him about her prospects for a new job with an excellent company known for its solid financial state and superior treatment of employees. Topit asks Emp for a job recommendation to support her application to the company. Emp is also excited to hear about the job and although he does not mention his decision to Topit, applies for the same position. He writes the letter of recommendation, but puts off sending it until right before the deadline for completion.

As the rumor about Endrun's failure unfolds, Endrun Corporation's stock becomes nearly worthless, which affects both Infoboom's and TechComplicated's businesses.

Case Context Questions

1. Can you locate all the legal issues within the case situation?
2. Can you point out all the ethical issues?
3. How can each one of the actors in the contextual setting be legally characterized? Is there more than one potential characterization for some of the actors?
4. Are LeaseUs Leases' actions legal? Ethical?
5. Is Meghan an agent with a fiduciary duty to TechComplicated?
6. What kinds of legal and ethical duties does Infoboom have to its employees who choose the option of taking Endrun stocks?
7. Does TechComplicated have a legal and/or ethical right to be paid for its work on the YRless web site? What are the potential legal outcomes?
8. What are the possible outcomes of this complicated contextual situation relative to each of the business entities and individuals participating?
9. Now that you have covered the material in the book, create a proposal for how you plan to approach the legal and ethical challenges you encounter or will encounter in your work in the future. Make sure that you reflect on all the areas of law and ethics covered in the text, noting benefits and detriments of each of the choices you make, and justifying them to your reader. Your proposal cannot provide you with a foolproof plan to avoid legal and ethical conflict but can prepare you for what may greet you in the future and can give you a better understanding of the consequences of choices you will make.

WORKS CITED

Ahn v. Rooney, Pace, Inc., (SD.N.Y. 1985).

American Law Institute. Discussion draft on section 2-103 (a)(17) of UCC Article 2, Feb. 14, 2000.

Apple Computers, Inc. v. Microsoft Corp., U.S. Court of Appeals (9th Cir. 1994).

Arledge v. Gulf Oil Corp., 578 F 2d 130 (5th Cir 1978).

Basic Books, Inc. v. Kinko's Graphics Corporation, 758 F. Supp. 1522.

Berger, Matt. "Verdict Delivers Blow to the DMCA." http://www. pcworld.com/news/article/0,aid,108040,00.asp. Feb. 25, 2003.

Berman, Paul Schiff. "The Globalization of Jurisdiction." *University of Pennsylvania Law Review 151* (Dec. 2002): 311–515.

Black's Law Dictionary. 5th ed. New York: West Publishing Co., 1979. p. 496.

Calamari, John D., and Joseph M. Perillo. *The Law of Contracts*. 4th ed. St. Paul: West, 1998.

Campbell v. Acuff-Rose Music, Inc., 114 S. Ct. 1164, U.S. (1994). Decided March 7, 1994.

Caprock Industries, Inc. v. Wood, 549 S W 2d 420 (Tex. Civ. App. 1977).

Castle Rock Entertainment, Inc. v. Carol Publishing Group, Inc., 47 USPQ 2d 1321 (2d Cir. 1998).

CCNV v. Reid, 490 U.S. 730 (June 5, 1989).

Clark, Gregory, and Stephen Doheny-Farina. "Public Discourse and Private Expression." *Written Communication 7* (1990): 456–81.

Cox v. Hickman, H.L. Cs. 268, Engl. Rep. 431 (1860).

Digital Millennium Copyright Act. U.S. Code. Title 17. Chapter 12. Sec. 1201.

Dombrowski, Paul. *Ethics in Technical Communication*. Needham Heights, Mass: Allyn and Bacon, 2000.

Dragga, Sam. "A Question of Ethics: Learning from Technical Communicators on the Job." *Technical Communication Quarterly 6* (1997): 161–178.

Dred Scott v. Sandford, 60 U.S. 393 (1:56).

Drennan v. Star Paving Co. 51 Cal 2d 409, 333 (Cal 1958) 186.

Eldred v. Ashcroft. 23 S. Ct. 769 U.S. 2003. Decided Jan. 15, 2003; 71 USLW 4052, 2003; Copr.L.Dec. P 28, 537, 65 USPQ 2d 1225; 3 Cal. Daily Op. Serv. 426, 2003; Daily Journal D.A.R. 512; 16 Fla. L. Weekly Fed. S 44.

Electronic Frontier Foundation. "EFF Whitepaper: Unintended Consequences: Three Years Under the DMCA." http://www.eff.org/IP/DMCA/20020503_dmca_consequences.html. Feb. 5, 2003.

Faber, Brenton. "Toward a Rhetoric of Change: Reconstructing Image and Narrative in Distressed Organizations." *Journal of Business and Technical Communication* 12 (1998): 217–237.

Fair Use Doctrine. Title 17. U.S.Code Sec. 107. 1978.

Ford v. Wisconsin Real Estate Bd. (Wisc. 1970).

Fujita, Anne K. "The Great Internet Panic: How Digitization Is Deforming Copyright Law." 2 J. Tech. L. and Policy 1 (1996).

Galler, Bernard A. *Software and Intellectual Property Protection*. Westport, Ct.: Quorum, 1995.

Gilliam v. American Broadcasting Co., Terry Gilliam, Graham Chapman, Terry Jones, Michael Palin, John Cleese, and Eric Idle, individually and collectively performing as the professional group known as "Monty Python," Plaintiffs-Appellants-Appellees, v. American Broadcasting Companies, Inc., Defendant-Appellee-Appellant Nos. 75-7693, 76-7023, Nos. 913, 1058 - September Term, 1975 United States Court of Appeals for the Second Circuit 538 F2d 14; 1976 U.S. App. LEXIS 8225; 192 USPQ (BNA) 1 April 13, 1976, Argued June 30, 1976, Decided.

Hayes, David L. "Liability of Online Service Providers: Part III." *Computer and Internet Lawyer* 9, no. 12 (December, 2002).

Herrington, TyAnna K. *Controlling Voices: Intellectual Property, Humanistic Studies, and the Internet*. Carbondale: Southern Illinois Univ. Press, 2001.

———. "Ethics and Graphic Design: Rhetorical Analysis of the Document Design in the *Report of the Bureau of Alcohol, Tobacco* and *Firearms Investigation of Vernon Wayne Howell* Also known as David Koresu" *IEEE* Transactions on Professional Communication, v. 38 n. 3 Sept. 1995. 151–157.

———. "The Interdependency of Fair Use and the First Amendment." *Computers and Composition Special Issue: Intellectual Property* 15, no. 2 (1998): 125–143.

———. "Who Owns My Work? The State of Work for Hire for Academics in Technical Communication." *Journal of Business and Technical Communication* 13, no. 2, April (1999).

———. "Work for Hire for Non-academic Creators." *Journal of Business and Technical Communication* 13, no. 4 (October 1999).

Home Owners Loan Corp. v. Thornbrush, 187 Okl 699, 106 P 2d 511 (1940).

Hopper v. Lennen & Mitchell, Inc., 146 F2d 364 (9th Cir 1944) 364.

Imfeld, Cassandra. "Playing Fair with Fair Use: The Digital Millennium Copyright Act's Impact on Encryption Researchers and Academicians." *Communication Law and Policy* 8 (Winter 2003): 111–139.

J.E.M. AG Supply, Inc. v. Pioneer Hi-Bred Intern., Inc., 122 Supreme Court Reporter 593, U.S. Iowa, 2001.

Kant, Immanuel. *Groundwork of the Metaphysic of Morals*. Translated by H. J. Paton. New York: Harper & Row, 1964.

Kuester, Jeffrey R. "Attorney Sites Can Avoid Violations of Ethics Rules." *The National Law Journal*. Reproduced online at www.computerbar.org/netethics/nlj.htm.

Lindey, Alexander. *Plagiarism and Originality*. Westport, Ct.: Greenwood, 1974.

Lucy v. Zehmer, 196 Va 493, 84 SE 2d 516 (1954).

Markel, Michael. "An Ethical Imperative for Technical Communicators." *IEEE Transactions on Professional Communication* 36 (1993): 81–86.

Martin v. Federal Life Ins. Co. 109 III App 3d 596, 65 III Dec. 143, 440 NE 2d 998 (1982).

Mattel, Inc. v. MCA Records, Inc., 296 F3d 894 (9th Cir 2002).

Mears v. Nationwide Mut. Ins. Co., 91 F3d 1118 (8th Cir 1996).

Meehan v. Valentine, 145 U.S. 611, 12 S. Ct. 972, 36 L.Ed. 835 (1892).

Moye, John E. *The Law of Business Organizations*. 5th ed. St. Paul: West, 1999.

Neitzke, Frederic William. *A Software Law Primer*. New York: Van Nostrand, 1984.

New York Times Co., Inc. v. Tasini, 121 S. Ct. 2381 U.S., 2001. Decided June 25, 2001.

Nguyen, Xuan-Thao N. "The Digital Trademark Right: A Troubling New Extraterritorial Reach of United States Law." 81 *North Carolina Lew Review*, Jan. 2003: 483–562.

1976 Copyright Act. U.S. Code. Title 17. 1982.

Raquel v. Education Management Corp., 96 F3d 171, CA3 (Pa.), 1999. Decided Nov. 9, 1999. United States Court of Appeals, Third Circuit.

Raquel, a Partnership v. Education Management Corpporation; Art Institutes International, Geffen Records Inc.; Nirvana, a partnership; Elias/Savion, Inc.; Philip Elias, individually Raquel, Appellant. No. 98-3321. Argued April 5, 1999. Decided Nov. 9, 1999.

Reuschlein, Harold Gill, and William A. Gregory. *The Law of Agency*. St. Paul: West, 1979.

Restatement 2d of Agency.

Restatement 2d on Contracts.

Revised Uniform Limited Partnership Act.

Ribstein, Larry E. *Unincorporated Business Entities*. Cincinnati: Anderson Publishing Co., 2000.

Salinger v. Random House, Inc., 811 F2d 90. U.S. Ct. of Appeals (2d Cir 1987).

Samuelson, Pamela. "Anticircumvention Rules: Threat to Science." *Science* 293, no. 2028 (Sept. 14, 2001).

Savage, Gerald. "Redefining the Responsibilites of Teachers and the Social Position of the Technical Communicator." *Technical Communication Quarterly* 5 (1996): 309–327.

Shetland Times v. Wills, settled. http://www.netlitigation.com/netlitigation/cases/shetland.htm.

Sony v. Universal Studios, Inc., 464 U.S. 417, 451 (1984).

Ticketmaster Corp., et al. v. Tickets.Com, Inc., U.S. District Court, Central District of California, March 27, 2000.

Uniform Commercial Code. UCC §2 = 207.

Uniform Partnership Act. UPA §6 (1).

United States v. Dimitry Sklyarov, 1, *United States v. Dmitry Sklyarov*, No. 5 01 257 (July 7, 2001). Also available at http://www.eff.org/IP/DMCA/US_v_ Elcomsoft/20010707_complaint.html).

United States v. Ronitti, 363 F2d 662 (9th Cir 1966).

US Const, Art. I, Sec. 8, cl. 8.

Walzer, Arthur. "The Ethics of False Implicature in Technical and Professional Writing Courses." *Journal of Technical Writing and Communication.* 19, no. 2 (1989): 149–160.

Washington Post v. Total News, Inc., settled. http://www.law.gwu.edu/facweb/claw/WPvTNsettl.htm.

Work for Hire. U.S.Code. Title 17. Chapter 1. Sec. 101.

INDEX

abuse of power, 15
acceptance, 4, 53, 56–59, 122; effectiveness, 59
accord and satisfaction, 81
accounts stated, 82
adhesion contracts, 72
agency, 2, 14, 26–29, 93, 115, 123, 126, 127, 131; kinds of agents, 29
agency relationship, creating a, 29, 122; invalidation of, 35; mutual duties, 29
agent's authority, 34–36
agent's duties and obligations, 32–33
agreements, legally effective, 53
Ahn v. Rooney, Pace, Inc., 30
American Bar Association, 11
apparent authority, 35
Apple Computers, Inc. v. Microsoft Corp., 104
Arledge v. Gulf Oil Corp., 61
assignment and delegation of duties, 79–81
attorney-client privilege, 12
author, 96, 97
authorship, elements, 96; corporate legal fiction, 97
axis of power, test for ethics, 8, 9, 10, 11, 19, 21

bailments and agencies, 66
balancing test, 20, 21
Barbie Girl, 88
Basic Books, Inc. v. Kinko's Graphics Corporation, 106
battle of the forms, 57
Berman, Paul, 111
billable hours, 16
binding exchange, 54
breach of contract, 61, 69–73, 67, 121; damages, 79, 122; duty to mitigate damages, 79
business entity, 2, 26
business organizations, 41–51, 131
business relationship(s), 2, 25–52

Campbell v. Acuff-Rose Music, Inc., 108
capacity to contract, 72–73
Caprock Industries, Inc. v. Wood, 31

Castle Rock Entertainment, Inc. v. Carol Publishing Group, Inc., 108
categorical imperative, 8
Church of Scientology, 13
clean room defense, 103; test, 105–106
code sharing, 9
commissioned works, types, 99–100, 114
common practice, 7, 8, 19, 20
computer code, 14
Congress, 7
consideration, 4, 53, 61–62, 121
contracts, contract law, 2, 3, 4, 53–83, 93, 120, 127, 128, 131; acceptance by silence, 57; conditions, 65–66; creation, 54; duties, 53; elements of, 56; ending, 67–77; indefiniteness, 62–63; mistake, 64, 122; misunderstanding, 63–64; mutual mistake, 64; revocation of, 57; suit compensation, 78; unilateral mistake, 65
copyright, 4, 120, 125, 129; control, 95, 127; protection scope, 94, 114
copyright holders' benefits, 94
Copyright Term Extension Act, 96
corporation(s), 4, 11, 50–51, 122, 126; taxation, 50
Cox v. Hickman, 27

DMCA (Digital Millennium Copyright Act), 13, 104–105, 110
deep linking, 109–110
delegation of duties, 80
derivative works, 95
digital databases, 109
Digital Millennium Copyright Act (DMCA), 13, 104–105, 110
Doheny-Farina, Stephen, 11
Dombrowski, Paul, 9
Dred Scott decision, 11
Drennen v. Star Paving, Inc., 66
duress, contract, 71

EFF (Electronic Frontier Foundation), 111
Eldred v. Ashcroft, 96

Electronic Frontier Foundation (EFF), 111
electric agent, 30
email, 12, 15
employee, 25; elements of status, 98–99
employee, legal characterization, 3, 36–41, 115, 128, 129
ethics, 131, 7–23; calculus of, 9; ethical behavior, 2, 3
existence of prior art, 92
existing practice, 8
express statements, 34

Faber, Brent, 18
fair use, 106–109
fair use doctrine, 17
false implicature, 8
false inference, 8
Federal Rules of Civil Procedure, 16
fiduciary concern, 33
first amendment, 88, 108
fixity, 93
forseeability in contract breach, 79
free speech, 17, 108
frustration, 67, 68
Fugitive Slave Act, 11

general agent, 29
general partnership(s), 4, 42–45, 124
gift, 61
Gilliam v. American Broadcasting Co., 102
goodwill, 4, 87–89, 125
Gregory, Clark, 11

Hayes, David L., 111
Home Owners Loan Corp. v. Thornbrush, 31
Hopper v. Lennen & Mitchell, Inc., 75
hypothetical imperative, 8

illegality, contract, 70–71, 121
Imfeld, Cassandra, 111
implied authority, 34, 123
implied duties of agent, 34
impossibility, 67, 68

impracticability, 67, 68, 121
idea-expression dichotomy, 92, 120
in terrorem effect, 17
independent contractor, 31, 37, 115, 123, 128, 129; legal characterization, 3, 25, 36–41
infringement, 93, 102–103; defenses against, 104–106
inherent authority, 35
injunction, 77
intellectual property, 2, 4, 126, 131; protections, 86

J.E.M. AG Supply, Inc. v. Pioneer Hi-Bred Intern, Inc., 91
joint work, 100–102, 123, 128; effect of community property, 101

Kant, Immanuel, 8
Kuester, Jeffrey, 12

LLC (limited liability corporation), 4, 48–49; tax structure, 48; shareholders, 48; articles of incorporation, 49
law and ethics, characterized and contrasted, 7–23
legal precedent, 2
limited liability corporation(s) (LLC), 4, 48–49; shareholders 48; articles of incorporation, 49
limited monopoly, 93
limited partnership(s), 4, 45–47
Lucy v. Zehmer, 56

mailbox rule, 59–60
Markel, Mark, 9
Martin v. Federal Life Ins. Co., 61
master-student relationship, 30
Mattel Inc. v. MCA Records, Inc., 88–89
Mattel Toys, 88
Mears v. Nationwide Mut. Ins. Co., 56
Meehan v. Valentine, 27
meeting of the minds, contract, 63

merger, 93
misrepresentation, contract, 71

Nguyen, Xuan-Thao, 88
1909 copyright statute, 36
1976 Copyright Act 92, 107
novelty, 92

offer, 4, 53, 55–56; irrevocability, 59; knowledge of, 58; revocation, 58
organizational structures, 25–52
originality, 93

parole evidence rule, 75–77
partnership, 27, 129; ending a, 47–48
patent, 4, 91–92, 124, 125, 126
performance, 56, 59
personal use, 106–109
plagiarism, 17, 93
principals, kinds, 29; death of, 35
principal's duties, 33–4
promissory estoppel, 66–67, 78

quasi contract, 77

Raquel v. Education Management Corp., 101
ratified authority, 36
releases, 82
reliance, 59, 80
repudiation of contract, 74
reputation, 4, 13, 87–89
respondeat superior, 28
Restatement 2d of Agency, 27
restitution, 69
Reuschlein, Harold, 31
reverse engineering, 93, 124, 125
Rule, 11, 16

S corporation, 48
Salinger v. Random House, Inc., 108
Savage, Gerald, 19
scope of employment, 97–98, 128
security interest, 80

Seinfeld Aptitude Test, 108
servant, 30
sham, 61
shareholders, 11
sole proprietorship(s), 4, 41–42
spam, 15, 16
special agent, 30
special skills, 15
specific performance, 77–78
statute of frauds, 7
substantial performance, 67
substantial similarity, 93, 95, 103–104
substituted agreement, 81
Supreme Court, 96

tangibility and fixation, 93
telemarketer, 13
third party beneficiaries, 81
Ticketmaster Corp. et al. v. Tickets.Com, Inc., 110
Times v. Wills, 110
tort, 77
trade secret, 4, 89–91, 125; factors for deciding, 90; subjects of, 89
trademark, 4, 87–89, 125

UCC (Uniform Commercial Code), 30, 54, 57, 75
unconscionability, in contracts, 71–72
Uniform Commercial Code (UCC), 30, 54, 57, 75
unique artistic talents, 15
US Constitution, 88
US Patent and Trademark Office, 88, 91
United States v. Dimitry Sklyarov, 110–111
United States v. Ronitti, 30
universalization, 7, 8, 19, 20
utilitarian test for ethics, 7, 9, 19

Walzer, Arthur, 8
work-for-hire doctrine, 36, 95, 97–100, 114, 127